MARINE PHOTOSYNTHESIS

WITH SPECIAL EMPHASIS ON THE ECOLOGICAL ASPECTS

FURTHER TITLES IN THIS SERIES

1 J.L. MERO
THE MINERAL RESOURCES OF THE SEA

2 L.M. FOMIN
THE DYNAMIC METHOD IN OCEANOGRAPHY

3 E.J.F. WOOD
MICROBIOLOGY OF OCEANS AND ESTUARIES

4 G. NEUMANN
OCEAN CURRENTS

5 N.G. JERLOV
OPTICAL OCEANOGRAPHY

6 V. VACQUIER
GEOMAGNETISM IN MARINE GEOLOGY

7 W.J. WALLACE
THE DEVELOPMENT OF THE CHLORINITY/SALINITY CONCEPT IN OCEANOGRAPHY

8 E. LISITZIN
SEA-LEVEL CHANGES

9 R.H. PARKER
THE STUDY OF BENTHIC COMMUNITIES: A MODEL AND A REVIEW

10 J.C.J. NIHOUL
MODELLING OF MARINE SYSTEMS

11 O.I. MAMAYEV
TEMPERATURE–SALINITY ANALYSIS OF OCEAN WATERS

12 E.J.F. WOOD and R.E. JOHANNES
TROPICAL MARINE POLLUTION

Elsevier Oceanography Series, 13

MARINE PHOTOSYNTHESIS

WITH SPECIAL EMPHASIS ON THE ECOLOGICAL ASPECTS

by

E. STEEMANN NIELSEN

Freshwater Biological Laboratory,
University of Copenhagen,
Hillerød, Denmark

ELSEVIER SCIENTIFIC PUBLISHING COMPANY

Amsterdam — Oxford — New York 1975

ELSEVIER SCIENTIFIC PUBLISHING COMPANY
335 Jan van Galenstraat
P.O. Box 211, Amsterdam, The Netherlands

AMERICAN ELSEVIER PUBLISHING COMPANY, INC.
52 Vanderbilt Avenue
New York, New York 10017

Library of Congress Card Number: 74-29691

ISBN: 0-444-41320-0

With 64 illustrations and 9 tables

Printed in The Netherlands

PREFACE

It is impossible today to present a relatively short book on photosynthesis in such a way that it can be regarded as exhaustive. This is the case even if the book has to cover marine photosynthesis only. A multitude of details are now known concerning this fundamental process. However, a considerable number of these details are, in fact, indispensable for the understanding of even the main trends of photosynthesis. It has been necessary, therefore, to discuss several of the details to some degree.

The present book is centred around the ecological problems related to marine photosynthesis; but the literature covering even this special field is so large, that an exhaustive review was not considered feasible. It has been found practical only to present relatively few references, in order to make the book readable. This is especially the case for the first chapter which deals with more general aspects. Here it is easy to find the relevant literature elsewhere; cf. "the Annual Review of Plant Physiology" which has been published every year since 1950. With a few exceptions, only recent references are presented in the first five chapters. However, it is the author's hope, that nevertheless sufficient references are presented to make it possible for the reader to penetrate further into the various subjects.

It has been necessary, in several instances, to refer to articles describing experiments in which freshwater plants have been used. It is a matter of course that the marine and the freshwater plants in many, if not in most respects, behave identically.

It is hoped that both students and specialists in biological oceanography will be able to read the book with profit. Measurements, e.g., of the primary production in the sea are made by workers from various scientific sectors. Specialists in plant physiology have not been the most numerous contributors.

Amongst other things, the book presents a plant physiologist's personal experiences obtained during more than forty years of work with marine primary production and photosynthesis. It is therefore perhaps forgivable that a relatively considerable part of the book refers to the author's own investigations. The reader must decide if this is a drawback or an advantage. It is further obvious that the planktonic algae occupy by far the largest space. Although this, of course, is partly due to the author's special interests, it must be admitted in justice that these algae have played by far the major role in the investigation of marine photosynthesis.

The various techniques used for measuring the rate of photosynthesis, including the rate of primary production, are only briefly discussed. For an exhaustive treatment, it is possible to refer to several books covering this special matter, e.g., Vollenweider (1969, second edition 1974).

Several colleagues have kindly read parts of the book in manuscript form. This has been a great help. I thank professor N. Jerlov, professor E.G. Jørgensen and Mr. S. Wium-Andersen.

E. STEEMANN NIELSEN

CONTENTS

CHAPTER 1

INTRODUCTION — PHOTOSYNTHESIS AND THE ORIGIN OF LIFE IN THE SEA

OUR GLOBE BEFORE THE EVOLUTION OF PHOTOSYNTHESIS

Photoautotrophic plants are now the only primary producers of organic matter in nature. Even the chemoautotrophic bacteria are — although secondarily — dependent on the photoautotrophic plants. The reduced inorganic matter which the chemosynthetic bacteria oxidize and whereby they obtain the necessary energy for the synthesis of organic matter, is originally produced due to the activity of photoautotrophic plants. All organisms now living on our globe depend on the organic matter produced by the photoautotrophic plants. This includes all animals, but also all heterotrophic plants, such as fungi, and the non-photoautotrophic bacteria.

It could therefore seem appropriate to expect that the first organisms on our globe had been photoautotrophic plants. We know that this is by no means the case. Furthermore, long, long before the appearance of the first primitive organisms on the earth, organic matter had been found in the oceans, although it very likely was first produced outside the oceans on the solid surface of the earth.

During the first milliards (= 10^9; in the U.S.A. billions) of years of the existence of our globe, the atmosphere was a reducing one. Besides other gasses in small concentrations, methane and ammonia were found in high concentrations. On the other hand, no free oxygen was found. This permitted the ultraviolet radiation to penetrate the atmosphere. Oxygen absorbs ultraviolet light to a high degree. Under the prevailing reducing conditions, the energy from short-wave ultraviolet light, which does not penetrate well into water, but which reached the solid surface of our globe, was able to build organic matter by means of, e.g., methane and ammonia.

This was suggested already in 1922 as a theory by the Russian biochemist Oparin. His ideas, however, had very little influence on the contemporary science. About 10 years after World War II some other scientists reconsidered his theory for the first time. Calvin (1965) has presented a review. From experiments in the laboratory they were able to show that organic matter could, in fact, be made in the way suggested by Oparin. Furthermore, it was shown that not only simple organic substances, such as hexoses, were synthesized but also a lot of more complicated substances, such as the amino acids. In the presence of H_2S, a series of organic substances containing sulphur were produced. In fact, practically all substances which are to be found during the destruction of the complicated compounds present in living cells, could be shown also to be synthesized during experiments made under reducing conditions.

If ultraviolet light is used, the synthesis of organic matter is thus a simple photochemical process; the latter cannot by any means be compared with photosynthesis. The starting substances, such as methane, are in a reducing state. Therefore, ultraviolet light only provides the activation energy. In photosynthesis, on the other hand, a net accumulation of potential chemical energy takes place. The starting substances, carbon dioxide and water, are in an oxidized state.

As very likely there was then little possibility of decomposing the organic matter present in the oceans, we may expect that quite considerable concentrations of the various organic substances were found in the oceans during this period. As no oxygen was found in the atmosphere to reduce the ultraviolet light coming from the sun, quite a considerable amount of energy was present for the synthesis of the organic matter.

We can only guess how the first extremely primitive organisms were created in the ocean. In fact, we do not know anything about this very important event. However, "fossils" of various bacteria-like cells have been found in sediments having an age of about $2 \cdot 10^9$ years. The first organisms in the oceans must definitely have been very primitive heterotrophic organisms and because no oxygen was present, they were anaerobic.

THE START OF PHOTOSYNTHESIS

According to all evidence, the present photoautotrophic bacteria are close to the first photoautotrophic plants which evolved on the earth. Their cellular organisation seems to be more primitive than ordinary photoautotrophic plants. The most crucial difference from the latter is, however, the fact that they are not completely photoautotrophic. Besides light energy, they also need reduced substances such as molecular hydrogen and hydrogen sulphide. Finally, they are found only under anaerobic conditions, such as the conditions were at the early stage of our globe, when life started.

A major evolution took place in photosynthesis when the second photochemical system (cf. Chapter 2) was evolved. This enabled the organisms to liberate oxygen from water and thus made them independent of the presence of reducing substances. All photoautotrophic plants, with the exception of the bacteria, have this second photochemical system. It is, however, interesting to note that some seemingly primitive green algae are found which are able to change their photosynthesis under anaerobic conditions using the reducing substance hydrogen instead of the second photochemical system.

About 10^9 years ago, photoautotrophic organisms producing oxygen must have developed. In sediments being formed during this time chemical compounds containing a porphyrine ring system have been found which must have been formed during the destruction of either chlorophyll or cytochroms. Cytochroms, however, can only have developed after the beginning of the activity of photosynthesizing organisms. Cytochroms are only active in aerobic respiration and this kind of respiration implies the presence of molecular oxygen. According to all evidence, the only way in which such molecular oxygen can have been formed is by means of photosynthesis. It is thus without importance, whether the compounds mentioned above are remnants of chlorophyll or of cytochroms.

The start of the present-day ordinary photosynthesis liberating molecular oxygen is one of the major events in the development of our globe (cf. Echlin, 1966). Now the atmosphere could gradually change from reducing to oxidizing. It very likely took several hundreds of millions of years before the atmosphere had reached something like its present state, where about 21% is molecular oxygen. However, long before that time the synthesis of organic matter without the assistance of photoautotrophic plants had stopped. The appearance of oxygen in the atmosphere and above all of ozone in the ionosphere at the same time gradually decreased the amount of ultraviolet light reaching the surface of our globe.

Before and during the beginning of the period with photoautotrophic plants, life outside the oceans and freshwater lakes was impossible. At higher intensities ultraviolet light, especially the short-wave part of it, impedes the growth of living organisms. Terrestrial plants and animals were first possible, when the atmosphere had reached a state more or less resembling that of today.

The oceans are thus the birthplace of life and the only place where for hundreds of millions of years the organisms could live and evolution could take place. As freshwater lakes are always ephemeral, i.e. they have a very short time of existence compared with the geological periods, we may disregard them, when speaking of the early evolution. Thus, all the taxonomic groups which have an old origin came to existence in the sea. This is also clear according to all geological evidence. All the major algal groups developed just before or during the Cambrian period, from about 1000 to 500 million years before the present time. Blue-green algae, diatoms, red algae, brown algae and green algae are thus true marine plants, although they now may be found in freshwater and, for some taxonomic groups, also on land. Brown algae and red algae, however, have with very few exceptions stuck to the sea.

When the organisms were able to emerge from the sea and settle on land, a relatively rapid evolution took place. Thus, a lot of taxonomic plant groups developed. The last one was the group of Monocotyledoneae, belonging to the flowering plants. It is rather curious to note that only species of this group of terrestrial plants have abandoned the land and again gone into the sea and here become true marine plants, i.e. submerged species. As far as we know none of the other groups of terrestrial plants seem to have sent representatives out into the sea again at any time. This is true both for the most ancient groups of land plants, such as e.g. mosses and ferns, and for the group of Dicotyledoneae from which the Monocotyledoneae, the other group of flowering plants, developed.

CHAPTER 2

THE PROCESS OF PHOTOSYNTHESIS

HISTORICAL NOTE

It is not necessary in this book to present a detailed survey of the biophysics and biochemistry of photosynthesis. In the same way, we shall not give a proper historical outline of the developments in the study of the subject (cf. Loomis, 1960). Although some of the main trends of photosynthesis were elucidated already during the last part of the 18th century, and although further and important developments in the understanding of the process were made during the second part of the 19th century and the beginning of the present century, the real breakthrough in the knowledge was first made possible after World War II. The necessary techniques were then for the first time available. It is by no means peculiar that the process of respiration was elucidated, at least in all its main trends, during the nineteen-thirties. Ordinary, although refined, chemical analyses which were sufficient to investigate the many separate processes in respiration, were at hand. For the investigation of the biochemistry of photosynthesis, the availability of the isotope ^{14}C and the development of isotopic techniques combined with paper-chromatography were a necessary basis for tackling the problems. For the understanding of the biophysics of the process, new techniques, first available only after World War II. were also a necessary precondition.

In the present book, the ecological aspects of photosynthesis will first of all be discussed. However, many of these aspects are directly the results of either biochemical or biophysical processes in photosynthesis. They are thus an indispensable background for understanding the ecology, and we must, at least to a certain degree, also present the biochemistry and biophysics. As many excellent books have been written covering these subjects, only outlines are given here and we may, e.g., point to volume V (parts 1 and 2) in the Encyclopedia of Plant Physiology (1960). For ecologists wanting a relatively short outline of photosynthesis, the book by Fogg (1968) may especially be recommended. Plant physiology and ecology rather than molecular biology is the focus.

During photosynthesis, radiant energy is converted into chemical energy in such a way that back-reactions are avoided. Chemical compounds containing potential chemical energy are thus stored in the cells and can be used later either for the building-up of new structures in the plant or for producing the fuel for the process of respiration necessary for carrying out a number of life processes.

TWO KINDS OF PROCESSES IN PHOTOSYNTHESIS

During the beginning years of this century, it was discovered that photosynthesis includes two kinds of processes — photochemical and enzymatical. Fig. 1 shows two curves presenting the rate of photosynthesis at two temperatures as a function of increasing irradiance (cf. footnote, p. 13). We shall later discuss such curves in detail. Here we shall only state that the first part of the curves, i.e. at weak irradiances, has slopes presenting a straight line. Here the photochemical processes are limiting the overall process. Photochemical processes take place proportionally to irradiance. The initial slope showing a straight line indicates that the rate of photosynthesis is limited by photochemical processes. A further proof is the fact, also shown in Fig. 1, that the rate of photosynthesis at weak irradiance is independent of temperature. This is a characteristic of a photochemical process; compare, e.g., the photochemical process used in photography, where we do not take the temperature into consideration.

In Fig. 1 the curves bend at somewhat stronger irradiance, gradually becoming parallel with the abscissa. Now the rate of photosynthesis is limited by the rate of enzymatic reactions. We presuppose that carbon dioxide is present in a sufficient concentration. At stronger irradiance, the rate of photosynthesis increases with a factor of a little more than 2 when temperature increases by 10° C. This tells us that ordinary chemical processes are limiting the overall process. It is of course no definite proof that it is especially a case of enzymatic processes. However, enzymatic processes are chemical processes and a lot of facts tell us that the rates of all these processes in plants are governed by enzymes.

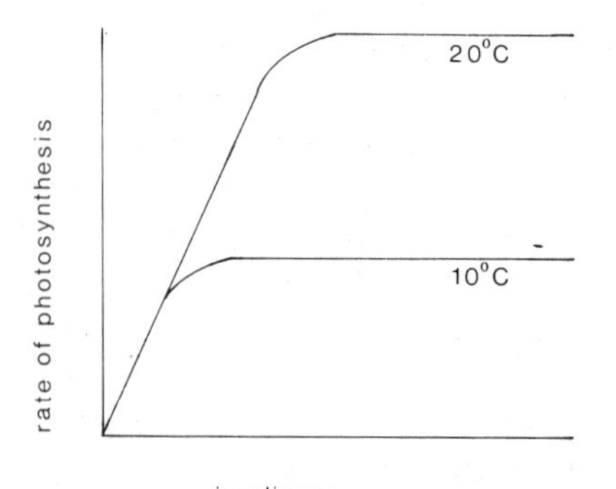

Fig. 1. The rate of photosynthesis as a function of irradiance at 10° and 20°C (schematic).

THE PHOTOCHEMICAL PROCESSES

Ordinary photoautotrophic plants are able to utilize light of between about 350 and about 700 nm. Photoautotrophic bacteria, on the other hand, are able to utilize light up to much longer wavelengths, some even up to more than 900 nm. When the ordinary plants are only able to utilize light up to 700 nm, this is partly due to the fact that in these plants none of the pigments active in photosynthesis are able to absorb light with wavelengths longer than about 700 nm. However, the upper boundary of photosynthesis at 700 nm is very likely not just an accident. As photosynthesis is a quantum process, a quantum has the same effect on photosynthesis independent of the size, if it only has sufficient energy (cf. p. 23). As the energy of the quanta decreases with increasing wavelength, it seems very likely, that the lower limit for energy in a quantum able to work in the photosynthesis of ordinary photoautotrophic plants is just the energy found in the red quanta, at about 700 nm. The photoautotrophic bacteria which, according to all evidence, demand practically the same number of quanta as the ordinary plants to assimilate a molecule of CO_2, utilize besides light energy also chemical energy. This may very likely be the cause of their ability to use quanta with less energy than those used in ordinary plants. At the same time, the pigments in these bacteria are able to absorb light at the longer wavelengths.

It must finally be mentioned that, according to Halldal (1968), an endozoic green alga (*Ostrobium* sp.) is able to utilize light up to about 750 nm. It lives deep inside a coral at extremely weak irradiances. It has a special pigment absorbing above 700 nm. As the alga must predominantly be heterotrophic, this perhaps makes it possible to utilize smaller quanta for photosynthesis than ordinary plants.

In photosynthesis, chlorophyll molecules — and molecules of other photosynthetic pigments — are able to absorb quanta and become excited. They have now more energy than the molecules in the ground state.

By ground state we understand the situation when the electrons in an atom tend to occupy the lowest energy levels. In a so-called excited state, on the other hand, an electron is displaced to an outer orbit. At least three different excited states are very likely found during photosynthesis. The first states, especially that produced by the blue, energy-rich quanta, have high energy, but they decay within an extremely short time to another state. Part of the energy is hereby liberated as heat.

Without going into details, we may state that the excited state with the least energy — the so-called triplet state — has, in comparison with the prior states, a much longer life (about 10^{-2} sec). Some of these activated triplet chlorophyll molecules (as will be mentioned later only a small special part of the chlorophyll molecules) are able, by means of their relatively long life and their excess energy (31 kcal. per g molecule), to expel their excess energy as an electron which reduces a special receptor, ferredoxin. Reduced ferredoxin may thus be considered as the first relatively stable product of photosynthesis. Ferredoxin has a redox potential of about −430 mV and it has, therefore, the possibility of doing real work. The now extra electron in the ferredoxin — or a corresponding one — will ultimately reduce the chlorophyll molecule which, due to the loss of the electron, had been oxidized.

THE ELECTRON TRANSPORT

Because a reduced ferredoxin molecule contains much energy, it would be uneconomical if the electron was directly transferred to chlorophyll. It has also been shown that this is by no means the case. Instead, the transfer takes place in smaller jumps each using up a certain amount of the energy. In Fig. 2 one way of electron transport is shown, the so-called cyclic electron transport. As some of the energy is used during the transport for phosphorylation, the reaction is also called cyclic photophosphorylation (ADP + P → ATP). Cyclic electron transport takes place both in ordinary photoautotrophic plants and in photoautotrophic bacteria. In the latter it can be the only electron transport.

For making the biochemical part of photosynthesis possible (see below), it is, however, necessary that besides ATP, also reduced NADP, i.e., the coenzyme phospho-nicotinamide adenine dinucleotide becomes available. In the photoautotrophic bacteria energy is partly or exclusively used for producing $NADH_2$, which in these organisms is used instead of the related $NADPH_2$.

In ordinary plants another kind of electron transport, the so-called noncyclic electron transport, is used for this purpose; concerning the photosynthetic bacteria see below. Fig. 3 presents the principle of the transport. The electron from the ferredoxin — in fact two electrons from two molecules of reduced ferredoxin — is here transferred to NADP. At the same time $2\,H^+$ is also transferred to NADP, thus producing $NADPH_2$. The $2\,H^+$ is

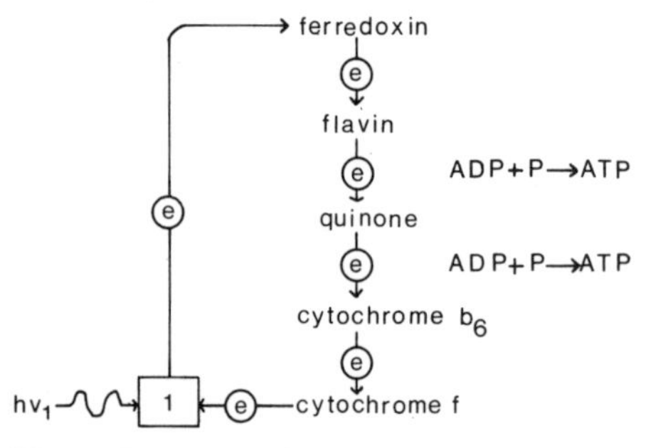

Fig. 2. Reaction scheme (model) for cyclic electron transport.

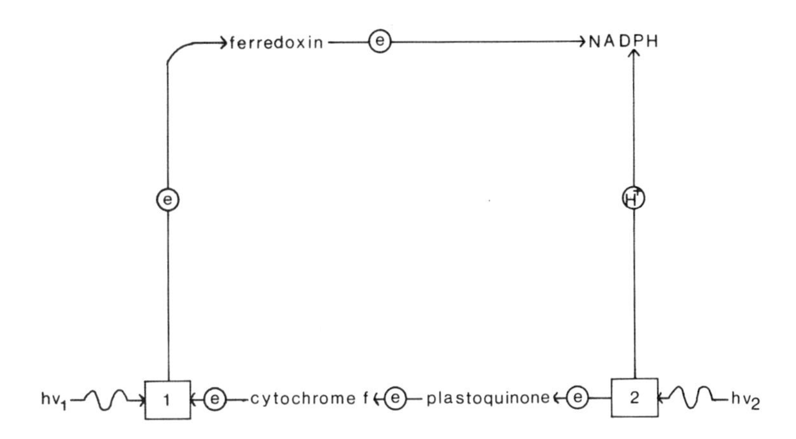

Fig. 3. Reaction scheme (model) for non-cyclic electron transport.

produced from water by means of another photochemical process (system 2) also involving excited special chlorophyll *a* molecules, but very likely of a slightly different kind from those in system 1. In system 2 water is split into H^+, O_2 and electrons ($2\,H_2O \rightarrow 4\,H^+ + O_2 + 4\,e$). The electrons are transferred in jumps via electron carriers to the oxidized chlorophyll molecules of system 1. Most of the energy of the electrons expelled from the chlorophyll molecules of system 2 is very likely used for the splitting of the water and these electrons probably return after the energy has been used to the molecules from which they were expelled.

In some photoautotrophic bacteria, molecular hydrogen can be used for the reduction of NAD. In this case no non-cyclic electron transport is necessary. Photoautotrophic bacteria, however, can also use e.g. thiosulphate as electron donor. In this case, the redox potential is not sufficiently low for the reduction of NAD. An extra energy supply is thus needed. We do not know how this energy is obtained, but we may guess that a non-cyclic electron transport takes place resembling to some extent that in the ordinary plants, however, without involving a splitting of water molecules.

For some years a discrepancy was found concerning the number of quanta used for the assimilation and of quanta necessary for assimilation of one molecule CO_2 in photosynthesis. A special study group under O. Warburg maintained that 3—4 quanta were enough. However, there is now general agreement that about 8 are necessary. All reliable measurements point in this direction, as do the schemes presented in Figs. 2 and 3. This means that about 40% of the light energy absorbed in photosynthesis is converted into chemical energy if red light near to 700 nm is used. Here we have the usable quanta with least energy. In blue light less than 30% is made into chemical energy.

THE PIGMENTS ACTIVE IN PHOTOSYNTHESIS

When discussing the photosynthesis of the various taxonomic groups of plants found in the sea (Chapter 6) we shall mention all the special photosynthetic pigments in detail. Here we shall restrict ourselves only to discussing the principal aspects.

The photosynthetic pigments are found in the chloroplast, if we disregard the bacteria which do not have such organelles. A chloroplast has a double-layered outer membrane and within this is a proteinaceous stroma. Embedded in the stroma are found double-membraned lamellae which in the higher plants and in some green algae are differentiated into thin stroma lamellae and thick grana lamellae. The latter are found in piles constituting the grana. The grana contain all the photosynthetic pigments. The chloroplasts in algae are built somewhat differently. For further details, the special literature must be consulted (e.g. Dodge, 1973).

Chlorophyll *a* is found in all the ordinary plants. This pigment has one absorption peak in the blue and one in the red part of the spectrum. The latter peak is at 678 in a living cell, but at 662 in ether solution. Chlorophyll *a* is incorporated in a definite structure in the grana. This structure is a condition *sine qua non* for the ability of the pigment to work in photosynthesis and it is also the reason why the absorption peaks in living plants are not exactly the same as in ether solution.

According to all probability, a special chlorophyll *a* modification is found. It is designated P 700, because a specially great decrease in absorption occurs upon illumination at 700 nm. This P 700 constitutes about 1/400 of the total amount of chlorophyll *a* and it seems to be able to "concentrate" the energy from a part of the ordinary chlorophyll *a* molecules by resonance. It should thus be the very place, from whence light energy is transferred into chemical energy by expulsion of an electron to ferredoxin (system 1).

Photoautotrophic bacteria do not possess chlorophyll *a*, but contain other chlorophylls absorbing up to much longer wavelengths than chlorophyll *a*. They have also been found to possess modifications in small concentrations which bleach upon illumination. The purple sulphur bacterium *Chromatium* has a special peak at 890 nm. This modification is thus called P 890 and it functions in all probability in the same way as does P 700 in ordinary photoautotrophic plants.

The only chlorophyll found in the blue-green algae is chlorophyll *a*. In practically all the other algae and in the higher plants, another chlorophyll is found as well; in the green algae and higher plants this is chlorophyll *b*. Also, other pigments are found in the chloroplasts, both biliproteins and carotenoids. We shall postpone most of the discussion of the function of all these pigments to Chapter 4, and here present only some general aspects.

Besides chlorophyll *a*, most — if not all — of the other photosynthetic pigments found in the chloroplasts are able to use absorbed light quanta indirectly for photosynthesis. However, as we shall see in Chapter 6, they are not all equally effective. None of the other pigments have, as does chlorophyll, a special kind of molecules able to expel electrons. However, several of the accessory pigments have been shown to be as effective in photosynthesis as chlorophyll *a*. The yellow pigment fucoxanthin, a carotenoid which is found in diatoms and brown algae, is able to absorb the light in the green part of the spectrum, where chlorophylls only absorb the light very slightly, but where fucoxanthin has its maximum absorption. The quantum yield (cf. p. 34) is the same at 550 nm, where fucoxanthin absorbs the light, as at 650 nm where all the light is absorbed by chlorophylls, especially chlorophyll *a*. This can hardly be explained in any other way than by its transfer of the energy absorbed by fucoxanthin to chlorophyll *a*. The most likely possibility is that the transfer of the energy is effected by resonance (cf. Duysens, 1951). However, this implies that the various pigment molecules must be closely juxtaposed in regular orientation. At the same time it is necessary that the receiving molecules must have an absorption band overlapping that of the molecules primarily absorbing the light energy. However, it is not safe to say categorically that the energy is transferred by resonance; at least one more possibility for the transfer cannot be excluded.

In contrast to fucoxanthin, certain other carotiniods found in photoautotrophic plants are only able to take part in photosynthesis to a slight degree. It is most likely that these pigments act primarily as traps for energy absorbed by other pigments such as the chlorophylls. If the rates of the enzymatic processes are too slow, more energy is absorbed than can be used in photosynthesis. We have therefore to expect that photooxidation would take place, destroying, i.a., a part of the photosynthetic mechanisms. If the surplus is trapped by carotinoids, photooxidation is prevented. Photooxidation and the counteracting thereof is discussed in more detail on p. 75.

In Chapter 10 chromatic adaptation will be discussed; this is of real importance for several algal species found at various depths.

THE PATH OF CARBON IN PHOTOSYNTHESIS

During the cyclic and non-cyclic electron transport, ATP and $NADPH_2$, or more correctly $NADPH + H^+$ (in photosynthetic bacteria $NADH + H^+$) are produced by means of the energy in the reduced ferredoxin. In the principal carbon reduction cycle (often called the Calvin cycle after its discoverer), organic matter, e.g. hexose sugar, is synthesized from CO_2 and hydrogen from $NADPH_2$. As the energy found in the reduced hydrogen-donor is not sufficient for this process, ATP supplies the rest of the energy. If glucose is the end product, we have the following equation:

$$12\ NADPH_2 + 18\ ATP + 6\ CO_2 \rightarrow C_6H_{12}O_6 + 18\ ADP + 12\ NADP + 6\ H_2O$$

However, before the reduction can take place, CO_2 is taken up — directly or indirectly — by an acceptor, ribulose diphosphate (RuDP), resulting in a compound with 6 carbon atoms. It is very unstable and splits into two molecules of PGA, phosphoglyceric acid. It is this compound which is reduced by means of ATP and $NADPH_2$, resulting in the formation of glyceraldehyde phosphate. We have now a compound which is able by means of a series of enzymes — but without any extra energy — to be transferred into many other substances having the same energy level, such as phosphate sugars, e.g. hexoses and pentoses. Ribulose monophosphate is thus also formed. By means of an ATP molecule it is transformed into ribulose diphosphate, the CO_2-acceptor in photosynthesis.

It is unnecessary to go into details concerning the Calvin cycle. These can be found in any textbook of biochemistry or in the book on photosynthesis by Fogg (1968). It is, however, important to note that some variations of the processes have been found both in higher plants and in algae. Hatch and Slack (1970) have published a review, the main points of which are presented below. In the unicellular green alga *Chlorella* grown at a high concentration of CO_2, it has been shown that for a time the concentration of the enzyme RuDP-carboxylase is too low to give rise to a sufficiently high carboxylation rate, if the algae are transferred to conditions with a low (normal) CO_2 concentration. There seems to be a close relationship between the Calvin cycle and a light-dependent release of CO_2, generally termed photorespiration. Evidence seems to have been provided that glycolate is the substrate for photorespiration, the physiological function of which is not really understood (cf. p. 51). Glycolate is derived from a phosphorylated intermediate of the Calvin cycle. It has been suggested that the process operates to prevent a total depletion of CO_2 within the chloroplasts. This is in accordance with the fact shown by Watt and Fogg (1966), that glycollic acid is liberated in quantities from *Chlorella* cells, if these are transferred from conditions of high carbon-dioxide concentration and low irradiance to low carbon-dioxide concentration and high irradiance.

In some higher plants, such as sugarcane and maize, a completely different process for CO_2-fixation has been found. It is termed the C_4-dicarboxylic acid pathway, in which CO_2 is taken up due to carboxylation of phosphoenol pyruvate. Oxaloacetate so formed is interconverted with pools of malate and aspartate. In some marine algae the C_4-dicarboxylic acid pathway has been observed to take place. However, only aspartate seems to be produced (Karekar and Joshi, 1973).

It was already mentioned that the light processes in photosynthesis take place in the grana of the chloroplasts. The enzymatic (dark) processes, on the other hand, take place in the stroma. Photosynthetic bacteria have no chloroplasts. Photosynthetic processes, however, are nevertheless centralized in very small organelles, called chromatophores. They are lamellar in structure.

CHAPTER 3

UNDERWATER DAYLIGHT

INTRODUCTION

The optics of the sea have been treated in another volume of this series — Jerlov (1968). It is highly recommended to consult this outstanding contribution which deals with all aspects of marine optics. In the present book, therefore, we shall by no means discuss all sides of this voluminous subject. On the other hand, certain aspects not so important for physicists must be stressed. Further, the variations of the irradiance[1] at the surface of the sea must of course be included in a discussion of submarine light if the effect of the irradiance on the rate of photosynthesis of marine plants is to be aimed at.

A physicist will normally measure light in energy units, such as Watts per area. This is quite obvious. It is the transfer of energy which is the focus of the physicist's interest. For a plant physiologist the problem is different. Photosynthesis is, as shown in Chapter 2, a quantum process. If a quantum can be used at all in this process, the energy content of the quantum is without importance.

INSTRUMENTATION

To overcome the important fact, mentioned above, Jerlov and Nygaard (1969) constructed an underwater irradiance meter[2] measuring the irradiance in quanta in the part of the spectrum, 350—700 nm, where photosynthesis in ordinary photoautotrophic plants takes place (cf. p. 7). This meter is easy to handle as it works exactly in the same way at both a strong and a weak irradiance. This is by no means true for all underwater irradiance meters, where it is necessary to introduce correction factors at stronger

[1] By irradiance, physicists understand the energy or number of quanta per second falling on a surface from all directions (see also Chapter 2). The term "illumination", which has been used by biologists (see e.g. Hutchinson, 1957) should be avoided, as the photopic sensitivity of the human eye is involved amongst other things (Jerlov, 1970). In order, like the physicists, to employ strict definitions of the quantities, the term light intensity is also avoided. Intensities can only be used in connection with a punctiform source of light.

[2] Similar meters made commercially are now on the market.

irradiances. The collector in Jerlov and Nygaard's meter, although not being a true cosine collector, is sufficiently correct to satisfy the requirements for which a plant physiologist will ordinarily ask.

We shall not go into detail concerning all the techniques of underwater light measurement. We shall just refer to Chapter 8 in Jerlov's book (1968). On the other hand, it is necessary to recommend biologists who are starting to measure light to contact a physicist familiar with measurements of light. Many pitfalls may turn up. This is especially the case if the distribution within the spectrum has to be investigated. This is necessary for the understanding of various problems in photosynthesis, especially if marine plants are to be investigated. As will be shown in Chapter 6, such a variety of photosynthetic plants is found, all with a very special absorption in the various parts of the spectrum.

Sophisticated meters exist: e.g., that constructed by Halldal which registers the distribution of quanta within the part of the spectrum of importance for photosynthesis. Also underwater irradiance meters using narrow-band filters — interference filters — can be mentioned. Meters using broad-band filters can be used with success for many biological purposes. However, they have the obvious defect that they most often gradually shift the optical centres when the meter is lowered into the sea.

Instead of measuring irradiance in absolute units at the various depths, several workers have measured it relative to that just below the surface. Working in such a way, there is no need to calibrate the meter in absolute units, a procedure which needs the assistance of a real specialist, who is not always available.

To measure submarine light is by no means a simple matter, not even if the correct instrument is at hand together with a specialist. The weather conditions are of major importance. The ship may be rolling so extensively, that work is more or less impossible. Drifting clouds changing the irradiance at the surface rapidly from one moment to another is a further complication. Measurements which demand a high precision should not be made under such conditions. Physicists will thus normally wait until the conditions for the measurements are favorable. This is not always possible for biologists, e.g. during a cruise where measurements of primary production in the sea are being made. Biologists are often forced to work at sea under rather adverse weather conditions. In biological work a high precision in light measurements is not always needed, because the natural variations in photosynthesis may be large.

THE INCIDENT LIGHT REACHING THE SURFACE OF THE SEA

Only a part of the solar radiation — corresponding roughly to the visible part — can be used for photosynthesis by the ordinary photoautotrophic

TABLE I

Solar radiation (direct + diffuse) according to Kimbal (1935)

Locality	Latitude	Annual total (kcal./cm^2)
Miami	25° 41′N	156
Washington	38° 56′N	124
New York	40° 46′N	100
Fairbanks	64° 55′N	76
Spitzbergen	79° 55′N	60

plants. We may estimate this part to be about 47% of the total (cf. Vollenweider, 1974). The intensity of solar radiation as received at the surface of the sea varies with latitude, season of the year, time of the day and the cloudiness.

Measurements of the annual totals of the incident light reaching the surface of the sea are presented in Table I. At 65° N (Fairbanks, Alaska) the annual total is about 50% of that at 26° N (Miami, Florida).

Whereas the seasonal variations in the tropics and subtropics are rather insignificant, they are very large at high latitudes, as shown in Table II. The average of the daily totals during the week that includes the winter solstice at Miami is 55% of that during the week that includes the summer solstice. At Fairbanks, just south of the Polar Circle, it is only 0.8%. At New York, 41° N, it is 21%.

With an overcast or partly overcast sky the radiation received is reduced compared with that on days with a clear sky. Daily continual measurements of irradiance were made during more than two years in Copenhagen, 56° N. Fig. 4 presents curves showing in relative units: (a) the average monthly curve; (b) the curve for the four brightest days of every month; and (c) the curve for the four darkest days of every month. During the four summer months, it was found for every month that on an average for 75% of the days the solar radiation was higher than 50% of that measured during the brightest days of the month; 94% of the days received more than 33% of the

TABLE II

Weekly averages of daily totals of solar radiation (direct + diffuse) received on a horizontal surface (cal./cm^2) according to Kimbal (1935)

Midweek date	Miami	New York	Fairbanks
March 22	456	284	221
June 21	514	435	522
September 20	476	299	120
December 20	283	93	4

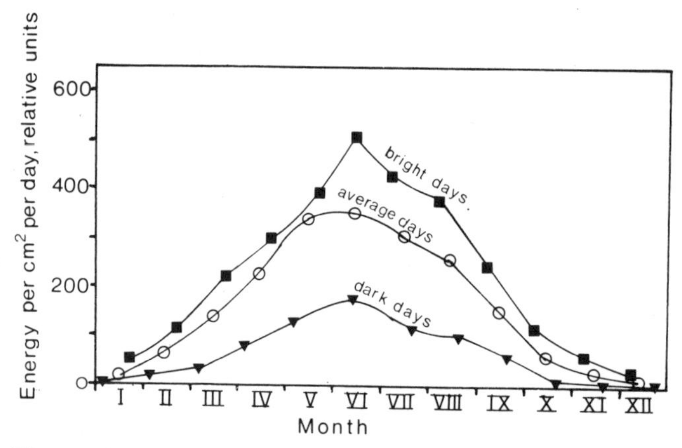

Fig. 4. The variation of the integrated daily irradiance throughout the year in Copenhagen (55° N) for bright, average and dark days. Light measurements by Romose (1940).

irradiance on the brightest days. This is important to note, because slighter variations at the surface influence the rate of primary production of phytoplankton relatively little during summer. This is discussed in detail on p. 104. When dealing with winter conditions at higher latitudes, the situation is of course quite different.

UNDERWATER LIGHT

At the very surface, part of the radiation is reflected and propagates back into the air. This part is dependent on solar elevation, wavelength, roughness of the sea (waves), and air bubbles found near the surface. We shall not go into any details (cf. Jerlov, 1968). The same is the case concerning the penetration and distribution of underwater radiant energy. However, as these properties are of major importance for the rate of photosynthesis going on in the water, we must present some of the most pertinent aspects.

The penetration of underwater radiant energy is determined both by absorption and scattering processes. We may distinguish between downward and upward irradiance, the latter being the result of scattering. By downward irradiance, which is by far the most important energy source for submarine photosynthesis, the physicists understand the radiant flux on an infinitesimal element of the upper face (0—180°) of a horizontal surface containing the point being considered, divided by the area of that element (see also the footnote on p. 13). Biologists call it the vertical "illumination" (should be irradiance). However, irradiance at a given depth may have other meanings besides. Biologists also speak of maximum "illumination" (irradiance), which at a certain depth is the irradiance on a recording surface set at such an angle, that the surface receives the maximum possible irradiance. Finally, the total "illumination" (irradiance) at a certain depth is that received by a point recorder from all directions.

It is obvious that total irradiance would be the ideal measure when measuring photosynthesis in small plankton algae and maximum irradiance would be the ideal measure, e.g., for solid, larger algae. Nevertheless, we are normally restricted to using vertical (= downward) irradiance, because this is measured by an ordinary underwater irradiance meter.

Due to absorption and scattering, the radiation is attenuated on its way downwards. The most convenient attenuation to use is the vertical attenuation and we define the vertical attenuation coefficient by:

$$E_z = E_0 \cdot e^{-K \cdot z}$$

where z is the depth in m; K the attenuation coefficient; e the basis of natural logarithms; E_0 the irradiance just below the surface; and E_z the irradiance on a horizontal plane at a depth of z m.

If the sun is low and a relatively large amount of radiation comes from the sky, z is by no means the real optical path of the radiation. The latter is instead, on the average about 1.2 z.

If we have to consider photosynthesis of terrestrial plants, we may imagine irradiance as having at least to some extent always the same wavelength composition. It is true that both the solar altitude and the cloudiness influence the relative importance of the irradiance in the various parts of the spectrum. Nevertheless, these variations are not constant, and they have therefore not given rise to the evolution of various types of chloroplasts. All chloroplasts in all species of terrestrial plants absorb light of different wavelength more or less in the same way.

In the sea the matter is quite different. Near the very surface we have of course a wavelength distribution such as that found on land. Below the surface and especially in the lower part of the photic zone (cf. p. 20), we may observe extremely varied conditions concerning wavelength distribution. Both the depth and the optical type of water — especially due to oceanic or coastal conditions — are of importance.

Fig. 5 presents, for a clear oceanic water, the complete spectrum of down-

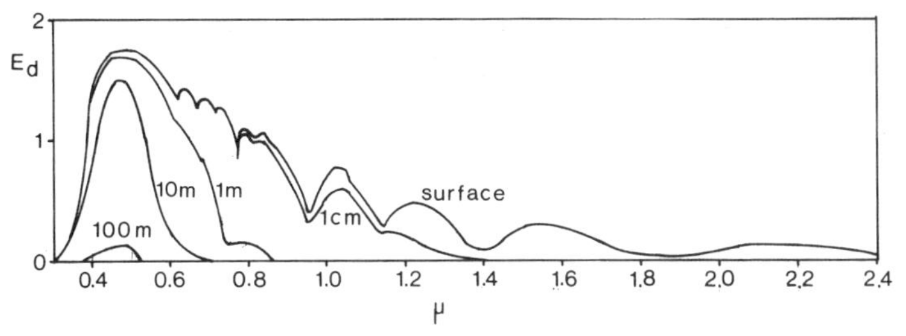

Fig. 5. The complete spectrum of downward irradiance in the ocean (after Jerlov, 1968).

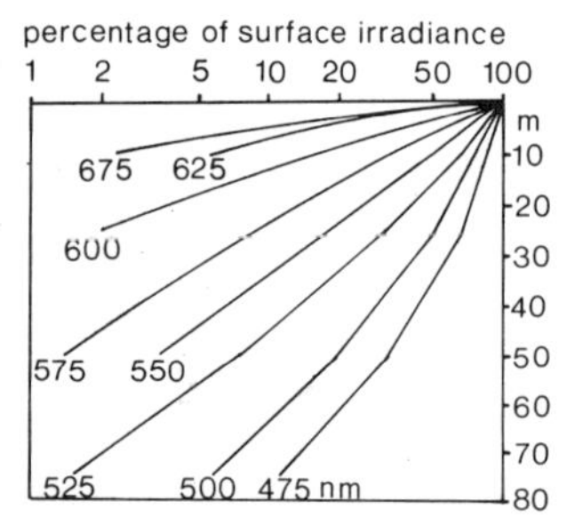

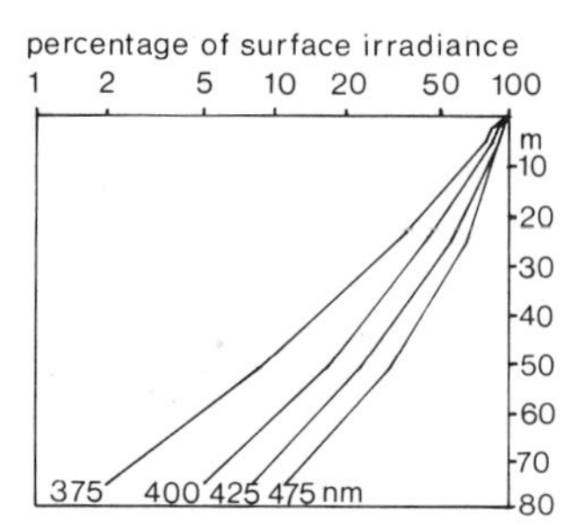

Fig. 6. Depth profiles of downward irradiance in the eastern Mediterranean (after Jerlov, 1968).

wards irradiance from sun and sky (300—2,400 nm = 0.3—2.4 μ). About half of the incident energy — the irradiance above 750 nm — is converted into heat in the upper water layer having a depth of less than 1 m. At a depth of 100 m the wavelength-selective attenuation mechanism has produced a narrow spectral range of nearly only blue light. Fig. 6 presents, for a station in the East Mediterranean with very clear water, depth profiles of downward irradiance in percent of surface irradiance for selected wavelengths in the spectral range 375—675 nm. Red light penetrates rather badly into the water. At a depth of 5 m we find only 10% of the surface radiation of red light (675 nm). For green light of 550 nm, 10% is found at a depth of 35 m, whereas the most penetrating blue light (475 nm) is reduced to 10% only at a depth of 82 m.

Polarization of scattered light is not dealt with in the present book (cf. Jerlov, 1968). Nevertheless, a possible effect on photosynthesis is discussed in Chapter 9.

OPTICAL TYPES OF SEAWATER

As already mentioned seawater may differ very much with respect to wavelength-selective attenuation. Jerlov (1951) described three optical types of oceanic water and nine types of coastal water. Fig. 7A presents the percentage of surface-downward irradiance (350—700 nm) as a function of depth (m) in the different water types. Irradiance was measured in energy. Fig. 7B presents the transmittance (in percent) per meter of downward irradiance in the surface layer for the various optical water types. Whereas blue light, as was already shown in Fig. 6, penetrates best in the clearest oceanic water (oceanic type I), in coastal water (types 1—5) green light has the highest transmittance. In coastal type 9 water, light in the yellow part of the spectrum penetrates the best. This tells us that the wavelength distribution is very different in the lower part of the photic zone from that near the surface. In some places we have practically only blue light, in others green

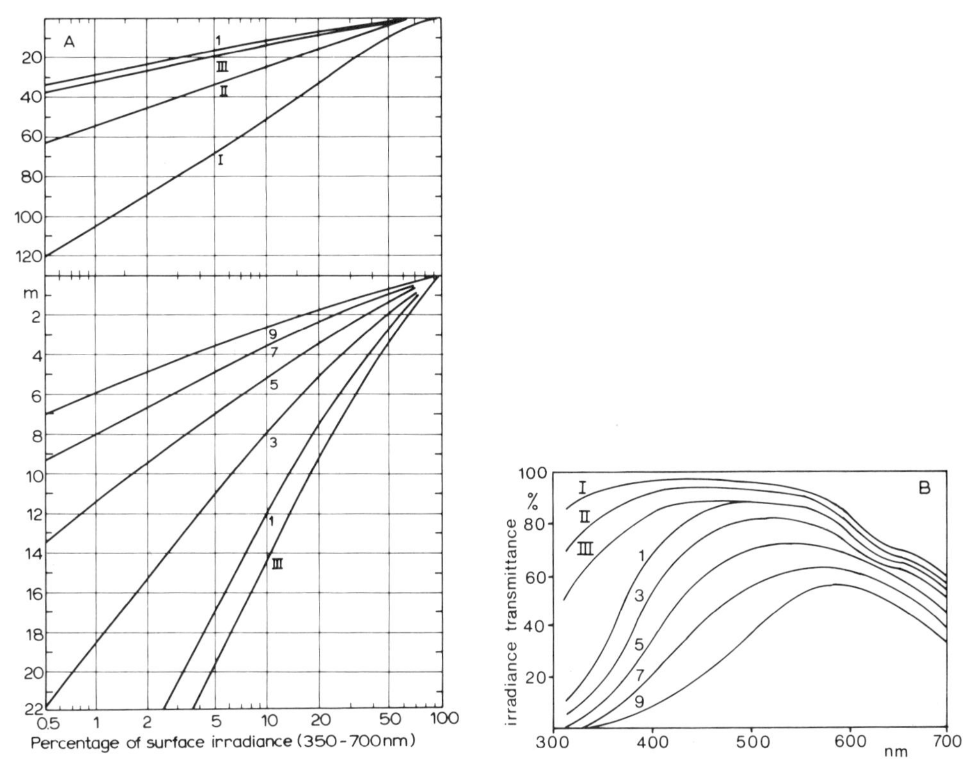

Fig. 7. A. Percentage of surface-downward irradiance (350—700 nm) as a function of depth (m) in defined water types (after Jerlov, 1970). B. Transmittance per meter of downward irradiance in the surface layer for defined water types (after Jerlov, 1968).

light and in some even preferably yellow light. This, of course, calls for the evolution of a variety of photoautotrophic plants having pigments absorbing light of different wavelengths differently. In coastal water (of type 5), for example, a composition of pigments absorbing practically only in the blue and the red part of the spectrum would be of little use for the plants living in the lower part of the photic zone.

The most clear oceanic water has practically the same attenuance as distilled water. This means that sea salts exert little influence on light attenuation with exception of that of short-wave ultraviolet light where the salts seem to cause a weak absorption. The attenuation of light of the various wavelengths, such as is shown for the clear water of the Eastern Mediterranean, is thus principally due to the water itself. Most of the attenuation is due to the absorption by the water molecules; some, however, is due to molecular scattering. It is only in the blue part of the spectrum that molecular scattering is of real importance. In turbid oceanic and coastal waters, molecular scattering is of relatively small importance. Here the scattering due to particles completely dominates. Particle scattering in the sea seems mostly to be practically independent of wavelength.

Particulate matter, which is found in very variable concentrations in the sea, often has a considerable influence on the attenuation of the light. Both absorption and scattering by the particles add to the attenuation. The particles may be of very widely differing kinds. Inorganic particles are found near the coast particularly, e.g., clay particles brought out with river water. Dead organic particles are numerous near the coast, but also in productive oceanic regions far from the continents. These particles are often much more numerous than the living organisms. In the open ocean the number of dead particles, however, is more or less a function of the number of living particles. Here we find that the transparency of the water is a function of the amount of plankton algae, and as will be shown in Chapter 13, therefore also of the rate of photosynthesis due to the plankton algae.

Numerous organic substances are found dissolved in seawater. Due to the dominant colour, a major group is collectively known as "yellow substance". It is partly produced in the sea as end products during the destruction, e.g., of carbohydrates. A considerable amount of yellow matter, however, is of continental origin, being carried out by the rivers. In the Baltic large amounts of humic matter come from Finland and the northern part of Sweden.

Yellow matter absorbs in the ultraviolet and blue part of the spectrum. Therefore, in coastal regions blue light usually is not sufficient for photosynthesis in the lower part of the photic zone. Kalle (1966) has presented a summary of all the aspects of yellow matter.

THE PHOTIC ZONE

General experience has shown that rates of photosynthesis, sufficient to compensate for the rates of respiration both during day and night in the tropics and the subtropics and also during summer at higher latitudes, take place down to a depth where approximately 1% of the surface light is found. This is the lower boundary of the photic, euphotic or photosynthetic zone or layer. Some authors prefer the sum of the green + blue light (Steemann Nielsen and Jensen, 1957). It is of no major importance, as we can only deal with an approximation. The lower limit of the photic zone is also called the compensation depth. Photosynthesis, of course, also takes place below the lower limit of the photic zone. Some indication is found that in the high Arctic, photoautotrophic algae can grow far below the depth at which 1% of the surface light is found (cf. p. 107).

SECCHI-DISC TRANSPARENCY

The Secchi disc is a white disc, normally with a diameter of 20 cm. It is lowered from the shaded side of the ship or boat into the water and the depth of disappearance is determined. The disc is then lowered a little more and subsequently raised again and the depth is determined at which the disc reappears. The mean of these two readings is the so-called Secchi-disc transparency.

Physicists seem to take Secchi-disc readings seriously only when speaking about visibility, which is beyond the scope of the present book. However, it would hardly be correct for biologists completely to ignore Secchi-disc readings. Many studies concerning primary production have been combined with them. The method is undoubtedly only a very approximate one. Nevertheless, the results are comparable in waters of more or less the same type.

In coastal waters of about type 3, it has been found by Atkins et al. (1954) that the depth of the euphotic layer is about 2.5 times the Secchi-disc transparency. The lower boundary of the euphotic layer is put, as mentioned above, at the depth where 1% of the light at the surface is found.

As the upward irradiance in the sea differs by at least between 2 and 5% from the downward irradiance, and as the disappearance of the disc is due to the identity of the light which it reflects and the upward irradiance, it is clear that the factor 2.5 just given cannot be correct for all water types.

CHAPTER 4

THE UNITS TO BE USED FOR THE IRRADIANCE IN MARINE PHOTOSYNTHESIS

The problem of how to measure and present irradiance has often been a difficult one for plant physiologists working with photosynthesis in the sea. As the light conditions here vary so much both quantitatively and qualitatively (see Chapter 3) and also because the various kinds of plants differ so much in how they absorb and utilize the light for photosynthesis, many problems arise. Such problems are generally much simpler when working with terrestrial plants.

Since photosynthesis is a photochemical process, the proper unit to use is quanta per second and per surface unit. This is especially obvious when working in monochromatic light. As will be shown in Chapter 6, a comparison of an absorption spectrum and an action spectrum for a plant species is only possible if quanta are used as the unit. A small quantum in the red part of the spectrum has the same effect in photosynthesis as a large quantum in the blue part of the spectrum, if it is absorbed by a photosynthetically fully active pigment.

The energy of a quantum in ergs is $h\nu$, where h = Planck's constant ($6.62 \cdot 10^{-18}$) and ν (the frequency of the radiation) = C/λ, where C = velocity of light ($3 \cdot 10^{8}$ m/sec) and λ = wavelength in nm.

If the illumination is presented in energy, physicists now practically always use the unit 1 watt/m^2; 1 watt = 1 joule/sec; 1 joule = 10^7 erg (erg = dyne · cm). Many biologists, however, still use the unit calorie, which is the energy necessary to raise the temperature of one g water by 1°C. We thus must present some conversion factors:

$$1 \text{ quantum/sec} = \frac{1987}{\lambda\ (\text{nm})} \cdot 10^{-19} \text{ watt}$$

1 watt = 14.3 cal./min

1 watt/cm^2 = 14.3 Ly/min

The unit Ly = Langley, 1 Ly = 1 cal./cm^2.

This multiplicity of units occurring in the literature makes the comparison of data often rather tedious. To make the situation still more complicated, moreover, some completely different units have been used by biologists, namely lux, klux (= 1000 lux) and ft-cdl (= foot-candles), 1 ft-cdl = 10.76 lux.

Lux values, which originate from the work with the human eye, have been chosen by several investigators for use in laboratory work with a definite light source. They have also sometimes been used for measuring daylight at the water surface. In both cases, it is more or less possible to convert lux values into real energy values. If a definite light source has been used, the conversion is rather good. For daylight, where the wavelength composition is by no means constant, the conversion factor is only an approximation. For noon sun + skylight in the range 380--720 nm, Strickland (1958) has presented the conversion 1 lux = $4.1 \cdot 10^{-7}$ W/cm^2. The light source must have at least some spectral resemblance with the standard candle on which the use of the lux unit is based. Therefore, for coloured light, the use of the lux as a unit is, in fact, meaningless. The same is also the case with submarine light. The future use of the term lux should now be abandoned.

In Chapter 9 experiments will be mentioned where the rate of photosynthesis is presented as a function of the irradiance from various light sources. Irradiance is given both in units of energy and in units of quanta. The most surprising fact is that the influence of measuring either in energy or in quanta is not too pronounced in most kinds of organisms. Only in the species of blue-green algae used was a conspicuous difference found (see Fig. 25). When in the following text irradiance is presented in $10^{15} \times$ quanta $\times$ cm^{-2} $\times$ sec^{-1}, both cm^{-2} and sec^{-1} are omitted.

CHAPTER 5

THE VARIOUS TYPES OF VEGETATION IN THE SEA

INTRODUCTION

Photosynthetic plants are found in all seas, but ordinarily only at depths or habitats where sufficient illumination is found, i.e. only in the photosynthetic zone, i.e. down to the depth where during 24 hours the rate of photosynthesis is higher than the rate of respiration. Photosynthetic plants — especially plankton algae and microbenthic algae — can also be found, however, in habitats where light in sufficient quantities for photosynthesis never will reach. Plankton algae may thus sink down below the photic zone and reach depths of more than 1,000 m (Lohmann, 1920). If this takes place in tropical seas with a lasting thermocline, the algae are lost for ever. If they are not eaten, they will die sooner or later due to lack of available organic matter to maintain life.

We may mention the following types of photoautotrophic vegetation in the sea:

a. phytoplankton
b. micropleuston
c. macropleuston
d. microbenthos
e. macrobenthos
f. epiphytes (both microscopic and macroscopic)

In special localities with anaerobic conditions we have particular vegetation of photosynthetic bacteria. In the sea this vegetation is benthic, living, e.g., at the surface of masses of decaying macroalgae. Planktonic vegetations of these bacteria, such as are found in various freshwater lakes, do not seem to be observed in the sea, where the combination of light and anaerobic conditions in the free water masses, if found at all, must at least be extremely rare.

PHYTOPLANKTON

In the sea, at least in all oceans, the phytoplankton is by far the most important vegetation. The amount of photosynthesis going on in the planktonic algae is many times greater than the sum of that due to all other kinds of marine plants together (cf. p. 119). When speaking of primary production in the sea, we usually think of the special one due to phytoplankton. The predominating role of the phytoplankton is by no means true for many shallow areas, where the organic production by other kinds of marine plants often exceeds that of the plankton algae. However, species attached to the bottom can only live around the oceans in a narrow fringe. As they must have sufficient light energy for their photosynthesis and because this is only found in the most clear water down to a depth of about 120 m, huge parts of the oceans with depths greater than 4,000 m are devoid of photosynthetic benthic plants, i.e. those which are attached to the bottom.

Although both micropleuston and macropleuston (see below) can also be found in the open ocean (the macropleuston predominately here), the rate of primary production by these plants is only low as compared with that of the phytoplankton.

The main bulk of phytoplanktonic algae is found in the taxonomic groups: diatoms, dinoflagellates, coccolithophorides, other flagellates, and blue-green algae. Species of green algae are only few in the sea and they are rarely found in large quantities. It is easy to understand that they avoid coastal waters; the green light, which dominates here in the lower part of the euphotic zone, is not suitable for photosynthesis in these algae (see Chapter 6). In ocean water, where blue light dominates in the lower part of the euphotic zone, the light quality can hardly be the cause of the more or less complete absence of green algae. Some species of green flagellates are indeed also found there, often even in considerable quantities.

Diatoms, dinoflagellates and coccolithophorides all have accessory pigments belonging to the carotenoids. This makes it possible for them to photosynthesize well at all wavelengths between 350 and 700 nm. Blue-green algae have biliproteins as accessory pigment. As is shown later, these pigments are formed in accordance with the quality of the light.

Planktonic algae are all small. This is a necessity in most parts of the sea in relation to the uptake of nutrients. Otherwise, the series of necessary compounds, which are found in relatively small concentrations, would hardly be taken up in sufficient quantities. The smaller the volume of an alga the larger is the relative alga surface; the size of this is important for the uptake of dissolved substances. Besides the nutrient salts, we also have to mention free CO_2, the carbon source for most plankton algae. As its concentration is only about $2 \cdot 10^{-5}$ moles and because much of this compound is needed for photosynthesis, the small size of the algae is an advantage. The constant sinking of plankton algae is also of decisive importance in this respect. Plankton algae sink into the ocean at a rate of about 5 m per 24 hours. For an alga having a diameter of say 10 μ this is a very rapid rate of sinking. It is about 50 μ per sec! The water around the cell is constantly and rapidly renewed.

At higher latitudes the thermocline disappears during winter and vertical mixing of the water masses may take place down to considerable depths. Specimens of algae, which have survived the winter conditions in the vertically mixed water layers in a more or less dormant state (or perhaps heterotrophically), will start again to act as photosynthetic algae if by chance they are found above the thermocline, when this is established during late spring.

MICROPLEUSTON

Only species of blue-green algae are found producing micropleuston in the sea. By pleuston we understand organisms floating on the surface of the sea. These algae are suspended deeper in the water during rough weather and thus act under such conditions exactly like the ordinary plankton algae. In calm weather they again ascend to the surface. Contrary to most freshwater species, the few marine species of blue-green algae forming micropleuston are predominantly found in oligotrophic areas — e.g. *Trichodesmium (Oscillatoria) thiebautii* in the Sargasso Sea. In the western part of the brackish Baltic Sea (also a relatively oligotrophic area), micropleuston due to a species of *Nodularia* is found during summer.

In very sheltered freshwater localities, photosynthetic algae may be found as neuston. Neuston organisms are attached to the very surface due to the powerful surface tension. Neuston algae seem never to be found in the sea. Bacteria, on the other hand, are present.

MACROPLEUSTON

After a storm in an unsheltered area where macrobenthic algae are found, a part of these algae may be detached; for a time they may thus be found floating at the surface. Such algae do not normally seem to grow and cannot be regarded as a true vegetation. However, in the central part of the North Atlantic, the so-called Sargasso Sea, real masses of some species of the brown algal genus *Sargassum* are found floating and growing at the surface. Gessner (1955, pp. 353—362) has presented a comprehensive review concerning the problems of the *Sargassum* in the Sargasso Sea. These algae — particularly *S. fluitans* and *S. natans* — seem to be endemic for the area. They are only able to propagate vegetatively. As the Sargasso Sea is one of the most oligotrophic parts of the oceans, the *Sargassum* species must very likely grow very slowly. The opposite has also been maintained, however. It would be very interesting to investigate the photosynthetic process in these plants. Judging from the behaviour in planktonic algae (cf. p. 52) we would perhaps suppose that considerable amounts of organic matter, only including C, H and O, but not N and P, are excreted.

The algae when seen floating are very conspicuous. The quantity, however, is by no means high, amounting per m^2 of the surface (recalculated according to Gessner, 1955) to only about 1.5 g fresh weight, corresponding with about 0.1 g C/m^2. This is about the same amount which is assimilated per day in the Sargasso Sea by the planktonic algae below 1 m^2 of the surface.

MICROBENTHOS

In coastal waters at shallow depths a flora of microalgae are found on the bottom. In the Kattegat, a sea situated between the North Sea and the Baltic, this vegetation is observed out to a depth of about 20 m (see p. 120). The irradiance at the bottom is about 5% of that at the surface. The quantitatively richest flora is at very shallow depth. Even on a shore exposed to the waves from the ocean, microbenthos is found (Steele and Baird, 1968).

Microbenthos algae attached to the sand grains near the shore may, during a storm, be buried deep in the sand masses. Grøntved (1962) has shown that in the Wadden Sea, in the southwestern part of the North Sea, living benthic microalgae may be found in large numbers down to at least 5—10 cm. However, such a distribution is only found in unsheltered places where, during a storm, the waves mix the sandmasses. The algae — mostly diatoms — are able to stand the dark conditions for several months. We do not know for certain how they survive. Heterotrophic life is one of the possibilities; a transformation to a dormant state is another. However, if algae which have been buried deep down in the sand, where light never reaches, are again brought into the light, photosynthesis will start very soon.

The depth of the photic zone in coastal sand is ordinarily only about 3 mm. In sheltered areas therefore, algae in numbers are not found very much deeper. Animals digging in the sand can of course transport some algae down to greater depths, even in sheltered areas.

Diatoms usually dominate completely at shallow depths. They excrete slime. In places exposed to waves they all have sufficient slime to be in close connection with the sand grains. Thus, if the sand masses are mixed during storms, the diatoms have to follow. However, in sheltered areas a considerable part of the diatoms is not in close connection with the sand grains or with the mud, if this is found instead of sand. A part of the diatoms is therefore easily stirred up in the water and will be a part of the plankton after a storm until, under quiet conditions, they again settle on the bottom. Green algae and blue-green algae are also found everywhere in the microbenthos. The same is the case for dinoflagellates and photoautotrophic flagellates.

In a sheltered Danish fjord, Grøntved (1960) has found between 4 and 11 millions of diatoms per cm^2. They were all found in the photic zone, i.e. down to a depth in the bottom sediment of 4 mm. On exposed areas in the Wadden Sea he found about 40 millions per cm^2. He only counted the algae down to a depth of 5 cm. As algae are found down to a depth of about 10 cm, the absolute amount below a surface of 1 cm^2 was, thus, much larger. In Chapter 14 we shall discuss the production rates of microbenthos.

MACROBENTHOS

Macrobenthic algae are found at shallow depths on most — but not at all — coasts. Further, some flowering plants, all belonging to the group of monocotyledons, are also present. Whereas the algae generally are attached to stones and other solid objects, the flowering plants have roots and rhizomes down in the bottom. It is thus a matter of course that these latter are only able to live in rather sheltered areas. Only very few species of flowering plants are really marine. For North European waters, in fact, only some few species belonging to the genus *Zostera* can be mentioned. In warmer areas species of *Thalassia* and *Posidonia* are found. In northern Europe there are also some few species of flowering plants which are restricted to brackish water, notably two species of the genus *Ruppia.* Finally, some freshwater plants are able to penetrate out into brackish water; *Potamogeton pectinatus* can live in water up to a salinity of about 18‰ S.

The fact that only a few species of higher plants are present in sheltered coastal waters does not mean that these plants are rare. In suitable areas huge populations of them are found.

Contrary to the small number of species of marine flowering plants, large numbers of benthic algal species are found in the sea. They are practically all truly marine. Some of them are able to penetrate into brackish water. Freshwater algae, on the other hand, ordinarily do not penetrate into brackish water. For North European waters only some few species belonging to the family *Characeae* (green algae) can be mentioned. The salinity must be rather low — up to about 7‰ S.

It is not the aim of the present book to go into details concerning the distribution of marine algae (e.g., Levring et al., 1969). Many ecological factors are of importance such as especially the tide, the exposure to waves, and the substrate. As we have to concentrate on photosynthesis, we have to mention particularly the depth and the optical water type. In the next chapter we shall see that it is the content of the various photosynthetic pigments in the algae which primarily determines the distribution of these plants in the lower part of the photic zone.

The three taxonomic groups: the red algae, the brown algae and the green algae are all typically marine. The two first-mentioned groups are even nearly exclusively marine, whereas the last group is more or less equally well at home in the sea and in freshwater.

Some algae are found in the supralittoral zone (also termed the adythalic zone), i.e. the zone above the ordinary high-water mark. They receive thus in most cases only a spray of water. They are nevertheless true water plants. In

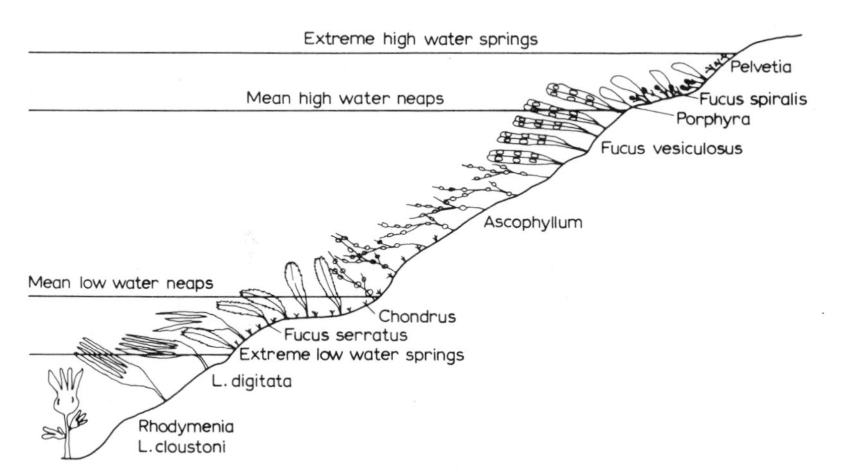

Fig. 8. Diagram illustrating typical zonation on a European rocky shore (according to Chapman in Levring et al., 1969).

the periods when they do not become moistened by the spray, they may dry out. They are able to tolerate both drying and the strong irradiance in such habitats. The same tolerance as is seen in the supralittoral zone is also more or less the case for the plants that live in the littoral zone (also called the hemidythalic zone), by which we understand the part of the sea bottom which alternately is covered or uncovered by the sea water. This can be due to the tide or in areas without any considerable tide, to the wind.

A flowering plant like *Zostera nana* does not dry out under natural conditions. It is able, by means of the roots, to take up water from the bottom, if for some time it is not covered by water. *Zostera nana*, however, is a true marine plant. It has no stomata, but photosynthesizes well both when it is submerged or outside the water. Some other plants living in some areas alternately covered and uncovered by water, must, on the other hand, be considered as landplants. As the most characteristic representative in northern Europe we can mention *Salicornia herbacea*. It has stomata like ordinary landplants. The major part of its photosynthesis takes place when it is out of the water.

Below the littoral zone, the sublittoral (or eudythalic) zone is found. The plants found here never become dry. The zone extends as far down as photosynthetic plants grow. In North European waters, the lower boundary of the sublittoral zone is at depths of about 25—40 m. In subtropical and tropical regions, even in some parts of the Mediterranean, clear oceanic water may be found near to the coast. The vegetation here penetrates down to much greater depths, e.g. to depths of rather more than 100 m.

As an example Fig. 8 is presented, illustrating a typical zonation on a European rocky shore. Above, we have used simple physical means for distinguishing the various zones. It must be mentioned that Lewis (1964) has used instead a zonation based on the vertical distribution of various organisms.

In the high Arctic, where the water masses are by no means very clear, algae have been found growing down to depths where the irradiance must be considerably less than that found at the lower boundary of the vegetation in temperate, tropical and subtropical waters (Lund, 1959; Wilce, 1967). In the Antarctic, Zaneveld (according to Wilce, 1967) has found benthic algae down to depths of 150—300 m. It must be a matter of fact that the algae growing at such considerable depths in very cold water have a different way of living from ordinary algae. We shall discuss this interesting enigma in Chapter 13.

CHAPTER 6

THE PIGMENTS FOUND IN THE VARIOUS TAXONOMIC GROUPS OF PLANTS AND THEIR IMPORTANCE FOR PHOTOSYNTHESIS

INTRODUCTION

As shown in Chapter 2, the overall process of photosynthesis is only limited by the photochemical processes at relatively weak irradiances. At strong irradiances, the enzymatical processes take over the role as the limiting factor. As the pigments take part in photochemical processes, it is necessary to work at a weak irradiance, if the influence of the pigments on photosynthesis is to be investigated.

The main pigments associated with photosynthesis are the chlorophylls (*a*, *b*, *c*, *d*) and special bacteriochlorophylls, the carotenoids carotene, lutein, fucoxanthin and peridinin, and the biliproteins phycoerythrin and phycocyanin.

In ordinary photosynthesis only those quanta are utilized which are absorbed by active chlorophyll *a* or another pigment able to transfer the absorbed energy to chlorophyll *a*. Normally the quanta absorbed by most pigments in the chloroplasts at low illumination rates are utilized in photosynthesis. As mentioned in Chapter 2, absorbed light energy may — very likely by resonance — be transferred from one pigment to another, ultimately arriving at the relatively few special chlorophyll *a* molecules which are able to expel electrons with a high energy content (cf. p. 10). Not all pigments found in the chloroplasts, however, are active in photosynthesis. Thus, several of the carotenoids seem to have another primary function (see p. 11). Not even quanta absorbed by chlorophyll *a* molecules at low illumination intensities are always utilized in photosynthesis. This may be due to a previous inactivation caused by a very high illumination rate (see p. 75) or it is a normal phenomenon as observed in many species of red algae (see p. 39).

If the influence of the wavelength is investigated, it is, as already mentioned, necessary to work at weak irradiances (Gabrielsen, 1960a). A low irradiance is, of course, automatically used if we select a very narrow wavelength band. It is in most cases in fact practically impossible under such conditions to produce a high irradiance.

It is usual to compare absorption curves (in percentage) and action spectra. The latter are obtained by measuring the photosynthetic yield for a given amount of quanta incident upon the plant over a range of wavelengths. The two curves are adjusted to coincide at a certain wavelength (see e.g. Fig. 11).

We may further use the term ϕ (= the quantum yield), which means the amount of CO_2 (in Mol) that is assimilated during the photosynthesis per amount of quanta (also in Mol) absorbed by the cells.

Whereas in higher plants, brown algae and red algae on the whole only include species belonging to the macrobenthos, green algae and blue-green algae both include large species and microalgae. Finally, all diatoms, peridinians and flagellates are unicellular (or in chains), both planktonic or attached to a substrate.

GREEN ALGAE

Fig. 9 shows the quantum yield as a function of the wavelength in the unicellular green alga *Chlorella pyrenoidosa* (according to Emerson and Lewis, 1943). In order to avoid the variable absorption of light in the various parts of the spectrum, a very dense suspension of the alga was used. In this way, even the green light, which is relatively poorly absorbed in a green alga, was completely absorbed.

The curve in Fig. 9 shows that the quantum yield is more or less the same in the spectral range 400—680 nm (for 685—700 nm see below), with the exception of that part of the spectrum (around 500 nm) where carotenoids play a role as absorbers of light. Fig. 10 shows the light absorption by a suspension of *Chlorella*. The minimum around 500 nm is due to the absorption by the carotenoids. The light absorbed by carotenoids in *Chlorella* is thus mostly rather badly used for photosynthesis.

The low quantum yield between 685 and about 700 nm (see Fig. 9) is due to a special phenomenon. By means of additional radiation by light of shorter wavelengths, the quantum yield in this region is enhanced (the Emerson effect).

Larger benthic green algae agree nicely with the unicellular alga *Chlorella*. In Fig. 11 the absorption spectrum and the action spectrum are shown for *Ulva taeniata* (according to Haxo and Blinks, 1950). We observe a decrease in the action spectrum at the wavelength where the carotenoids are partly responsible for the light absorption. This is the same as was shown for *Chlorella*.

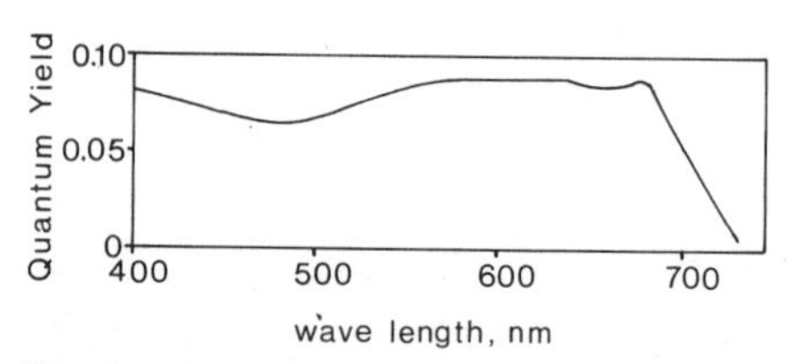

Fig. 9. Quantum yield as a function of wavelength in *Chlorella pyrenoidosa* (after Emerson and Lewis, 1943).

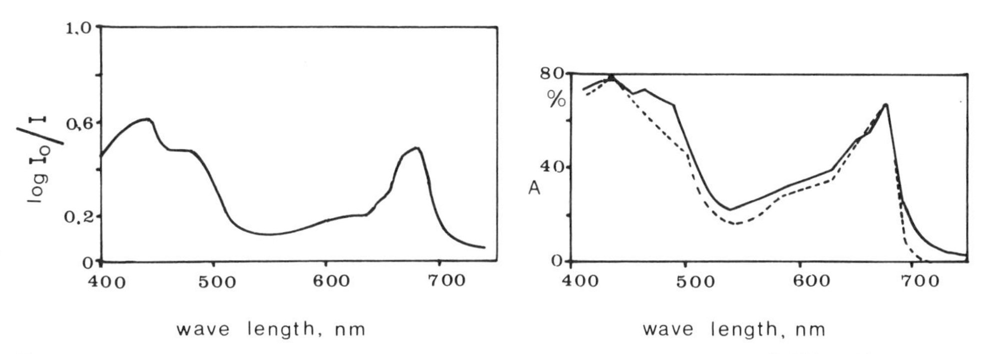

Fig. 10. Light absorption as a function of wavelength in a suspension of *Chlorella pyrenoidosa* (after Emerson and Lewis, 1943).

Fig. 11. Absorption spectrum (———) and action spectrum (- - - - - -) for *Ulva taeniata* (after Haxo and Blinks, 1950).

The curves in Fig. 11 demonstrated that green algae — at least those which, like *Ulva*, have a thin thallus with a rather low chlorophyll content per surface unit — are badly fitted to live in the lower part of the photic zone in coastal waters. As shown in Chapter 3, practically only green light is found here, and green light is badly absorbed by green algae. However, where oceanic water is found close to the coast, benthic green algae can be found down to relatively great depths (Levring et al., 1969). This is due to the blue light which in such waters is found in considerable quantity in the lower part of the photic zone. As seen from curve (——) in Fig. 11, blue light is absorbed vigorously by green algae due to the presence of chlorophyll.

The absence of green algae in the lower part of the euphotic zone in ordinary coastal regions is not absolute, however; some few species are nevertheless found there. *Chaetomorpha melagonium* preferably grows at rather great depth in the Kattegat, where practically only green light is found. However, the species has an extremely high concentration of chlorophyll. The name *melagonium* refers to its nearly black appearance. As mentioned above, a very high concentration of chlorophyll has the effect that green light, which is poorly absorbed by this pigment, nevertheless will be absorbed sufficiently for photosynthesis.

DIATOMS AND DINOFLAGELLATES

Fig. 12 shows the quantum yield as a function of the wavelength in the diatom *Navicula minima*. The yield is practically the same in the spectral range 520—680 nm, but shows, as for *Chlorella* (Fig. 9), a relative minimum at about 475—525 nm, where photosynthetically non-functioning carotenoids are found. Fig. 13 shows how the various pigments present in the

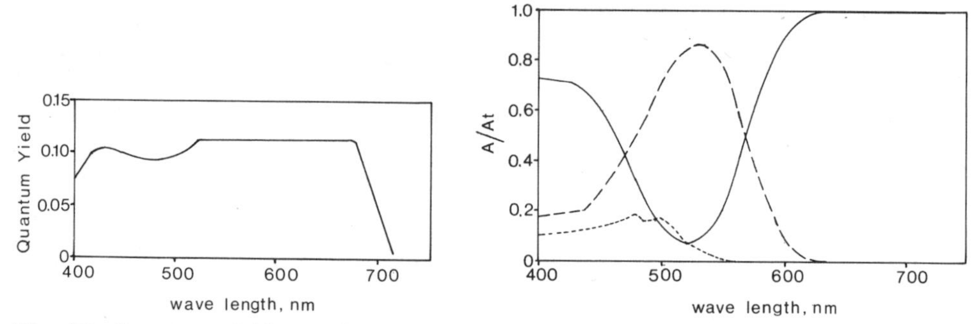

Fig. 12. Quantum yield as a function of wavelength in *Navicula minima* (after Tanada, 1951).

Fig. 13. The absorption of the various pigments in living cells of *Navicula minima* as part of the absorption of all pigments (A_t). Full line: chlorophyll $a + c$; dashed line: fucoxanthin; dotted line: the other carotenoids (after Tanada, 1951).

diatom absorb at the various wavelengths. In the green part of the spectrum, where the main pigment absorbing light is the carotenoid fucoxanthin, the quantum yield is the same as in that part of the spectrum where the main part of the light is absorbed either by chlorophyll *a* or *c*. This tells us that light absorbed by fucoxanthin and chlorophyll *c* is able to be transferred to chlorophyll *a* without any appreciable loss of energy. The remaining carotenoids other than fucoxanthin seem — as with the green algae — to be more or less unable to transfer the absorbed energy to chlorophyll *a*.

Fig. 12 shows that diatoms are equally able to live under all colour conditions in the sea. This is in agreement with their distribution. When making photosynthesis experiments in the laboratory in order to show the rate of photosynthesis as a function of increasing irradiance, even light sources having rather different colour distributions result in practically the same curves (cf. p. 55).

According to Halldal (1974), dinoflagellates have the accessory carotenoid peridinin, with absorption characteristics close to those of fucoxanthin. The ecology in respect to the quality of light seems thus to be about the same in diatoms and dinoflagellates.

Other groups of photoautotrophic flagellates are found in the sea, e.g. the brown coccolithophorides. These very likely behave more or less like the diatoms with respect to the utilization of light of different wavelengths. In the open ocean we even find green flagellates.

BROWN ALGAE

In this group we find species both with a thin and a thick thallus. This is of considerable importance if we are to compare the efficiency of photo-

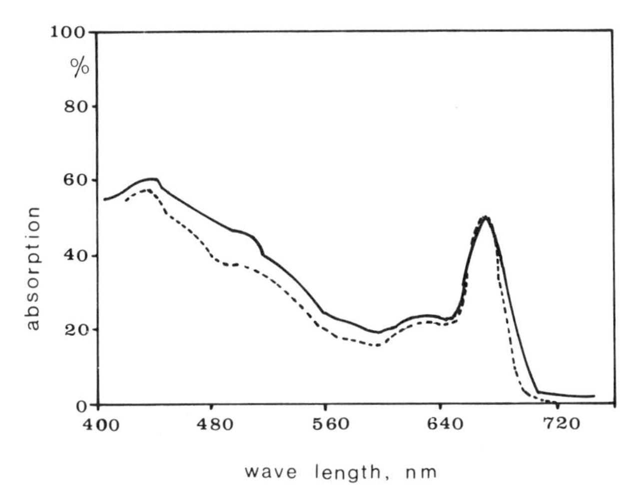

Fig. 14. Absorption spectrum (———) and action spectrum (- - - - - -) for *Coilodesme californica* (after Haxo and Blinks, 1950).

synthesis at various wavelengths. As shown in the curve in Fig. 9, quanta are utilized in photosynthesis of *Chlorella* both in the part of the spectrum where the pigments absorb poorly and in the parts where they absorb optimally. The suspension, however, had to be so dense, that all light was completely absorbed even in the green part of the spectrum, where the pigments had the lowest capacity for absorption. This is the background reason as to why thin and thick thalli do not give the same result either concerning the absorption spectrum or the action spectrum. This was emphasized by Gabrielsen (1960a).

Fig. 14 shows the absorption spectrum and the action spectrum for the brown alga *Coilodesme californica*, according to Haxo and Blinks (1950). The action spectrum and the absorption spectrum follow each other rather well if they are made to fit at 440 nm. Fig. 15 (according to Halldal, 1974) shows the corresponding curves for the brown alga *Laminaria saccharina*, which has considerably thicker thalli than *Coilodesme*. Unfortunately, the absorption spectrum was measured in a thinner piece than that used for measuring the action spectrum. The latter fact must be the reason why around 600 nm, where the minimum for absorption is found, the level of the action spectrum is found at a considerably higher level than that of the absorption spectrum. The two spectra were made to fit at 440 nm.

The curves presented in Fig. 14 and Fig. 15 show unambiguously that the carotenoid fucoxanthin in brown algae is able to transfer absorbed quanta to chlorophyll *a* in the same way as in diatoms.

Due to the presence of fucoxanthin, brown algae are able to penetrate down into the lower part of the photic zone in normal coastal waters. As, however, the absorption of green light is not particularly good, such as is the

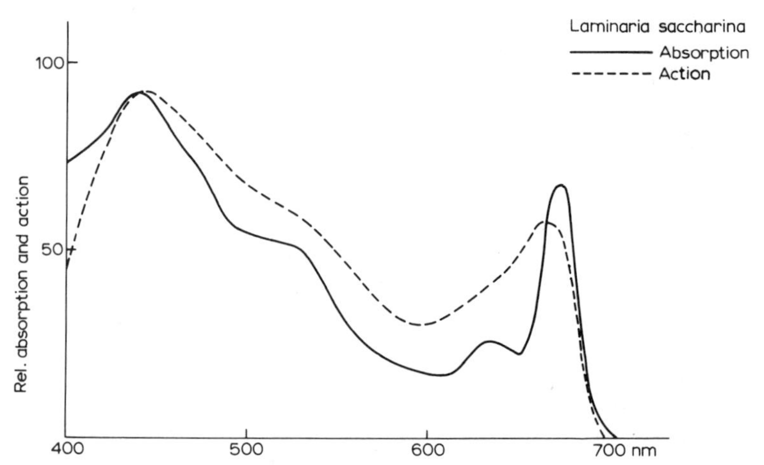

Fig. 15. Absorption spectrum and action spectrum in *Laminaria saccharina* (after Halldal, 1974).

case in red algae (see below) they are not ordinarily able to penetrate quite as far down as the latter.

RED ALGAE

This taxonomic group presents a rather peculiar situation concerning the pigments. Although for these algae also chlorophyll *a* is the very pigment which is able to transfer light energy to chemical energy, its own absorption of quanta ordinarily is of minor importance. Most red algae growing at shallow depths are even more or less brown, due to a considerable content of photosynthetically non-effective carotenoids. The importance of these carotenoids is very likely to protect against too high illumination intensities in the blue part of the spectrum.

In red algae some special photosynthetically active pigments are found. These are also characteristic for another taxonomic group, the blue-green algae. The pigments are phycobilins in connection with proteins (globulin). Phycobilins have four pyrrolic rings in a linear position and two kinds of them are found, the red phycoerythrins and the blue phycocyanins. They both seem to be present in all species of red algae, the latter, however, mostly in rather small concentrations.

In some few species of red algae, chlorophyll *d* is also found besides chlorophyll *a*.

For red algae, Levring (1947) was the first to investigate the effect of light in the various parts of the spectrum using a reliable technique. He investigated a whole series of species and found a much better utilization of the light for photosynthesis between 500 and 600 nm than at both shorter and longer wavelengths.

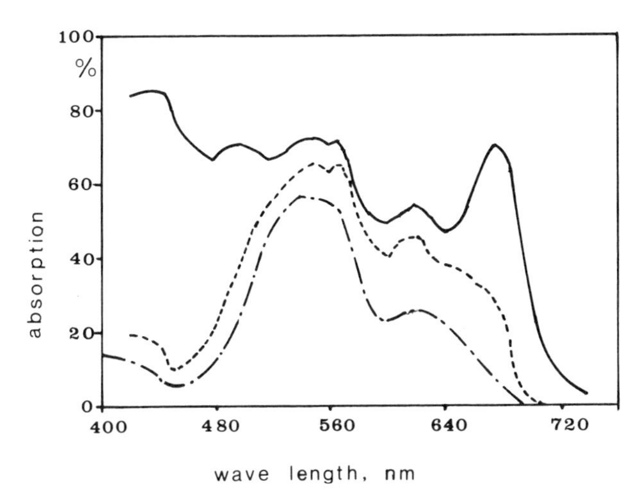

Fig. 16. Absorption spectrum (———) and action spectrum (- - - -) for *Porphyra naiadum*; (— · — ·) absorption spectrum for the phycobilins dissolved in water (after Haxo and Blinks, 1950).

This has been corroborated by Haxo and Blinks (1950) who compared the action spectrum and the absorption spectrum i.a. for the red alga *Porphyra naiadum* (Fig. 16). Here also the absorption is presented for the phycobilins dissolved in water. Compared with most other red algae, *Porphyra naiadum* has a rather high concentration of phycocyanin which has the absorption peak at about 615 nm, whereas the peak of phycoerythrin is in the green part of the spectrum (about 560 nm). The curves show that the phycobilins must be very active in photosynthesis. On the other hand, chlorophyll *a* and the carotenoids must be rather inactive in photosynthesis. In the part of the spectrum where these pigments are the only ones absorbing light, the photosynthetic yield is very low. In the part of the spectrum where, in red algae, the phycobilins are the main absorbers of the light, the quantum yield is only somewhat lower than the maximum yields found in green algae and brown algae (Table 4 in Gabrielsen, 1960b).

Yocum and Blinks (1958) have shown that it is possible in species of primitive red algae (Protoflorideae) to adapt plants to utilize light absorbed by chlorophyll *a*. This took place, if the algae were grown in blue or red light. However, if these plants were illuminated for some few hours with green light, this special adaptation disappeared.

BLUE-GREEN ALGAE

Fig. 17 presents the quantum yield as a function of the wavelength in a species of the genus *Chroococcus* (Emerson and Lewis, 1942). The measurements were made in the same way as with *Chlorella* and *Navicula* (Figs. 9

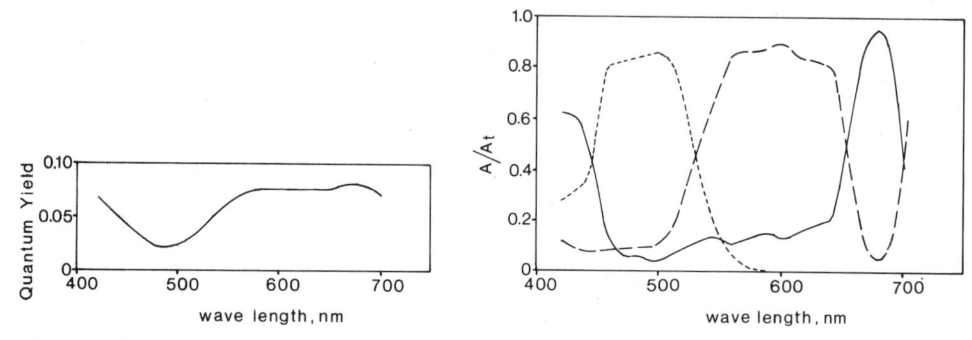

Fig. 17. Quantum yield as a function of wavelength in *Chroococcus* sp. (after Emerson and Lewis, 1942).

Fig. 18. Light absorption as a function of wavelength in various pigments of *Chroococcus* sp. Chlorophyll *a*: ———; phycocyanin: — —; carotenoids: - - - - - - (after Emerson and Lewis, 1942).

and 12). The maximum yield in the blue-green alga is nearly the same as that found for *Chlorella.*

Blue-green algae contain chlorophyll *a* and the phycobilins phycocyanin and phycoerythrin. In addition, carotenoids are also found. In the material of *Chroococcus* used for the experiments, phycocyanin was the only phycobilin. Fig. 18 shows the relative proportion of the light absorbed by the various pigments. It is obvious that light absorbed by phycocyanin must be able to be transferred to chlorophyll *a* with only a small loss. The drop in the quantum yield at wavelengths around 475 nm is due to the photosynthetically non-active carotenoids absorbing there.

Many species of blue-green algae adapt chromatically. At the surface they have the pigment phycocyanin primarily absorbing the light at wavelengths between 540 and 650 nm. Growing at depths where green light is predominant, they have instead phycoerythrin absorbing in the green part of the spectrum (with peaks e.g. around 565 nm). This has been especially investigated in freshwater species. Chromatic adaptation is discussed in Chapter 10.

HIGHER PLANTS

As far as is known to the author, no measurements concerning the comparison of action spectrum and absorption spectrum of higher marine plants have been made. However, we can hardly expect any difference in this respect between terrestrial and marine species. Although in principle the higher plants resemble those of green algae, we have to take into consideration that the leaves — e.g. those of *Zostera marina* — have rather high contents of chlorophyll per unit of leaf area. As mentioned on p. 35, this makes a relatively high absorption of green light possible. In coastal areas thus, *Zostera marina* may also grow at rather considerable depths — but not at the really great depths.

PHOTOSYNTHETIC BACTERIA

The amount of organic matter produced in the sea by marine photoautotrophic bacteria is exceedingly small compared with that produced by the "true" marine plants. On the other hand, these bacteria have played a considerable role in the understanding of the mechanisms of photosynthesis. A few remarks will be given below.

As shown by French (1938), *Rhodospirillum rubrum* is able to photosynthesize up to wavelengths around 900 nm. This organism is thus able to utilize considerably smaller quanta than ordinary plants. The contribution of chemical energy to the photosynthetic process and the presence of a special chlorophyll able to absorb at the long wavelengths make this possible (cf. p. 10). Special chlorophylls are found in the various groups of photosynthesizing bacteria. We may refer to Gabrielsen (1960b).

ECOLOGICAL CONSEQUENCES

For the plants growing at shallow depths the variation of pigments can hardly be considered to be very important. This is also clear because all the taxonomic groups are found abundantly there. As stressed in this chapter, the dependence on wavelength is a factor only of importance when irradiance is weak. During most times of the day, if we exclude the winter months at higher latitudes, the plants at shallow depths photosynthesize under more or less light-saturated conditions.

If any pigments should be especially mentioned in connection with the surface, it must be the non-photosynthetic carotenoids which have a protecting influence against too high illumination rates. It is typically seen in red algae, where the species living at shallow depths are often brown due to large concentrations of such carotenoids.

On the other hand, for algae living at greater depths the special composition of pigments must be considered to be of real importance.

Diatoms seem to be the taxonomic group most apt to grow under all the various light qualities. There seems to be no real difference between the effect on photosynthesis in the various parts of the spectrum. All other taxonomic groups are, at least to some extent, better suited to grow under one light quality than another.

In areas having optical coastal water types 1—9 (cf. p. 18), the light quality in the lower part of the photic zone is especially of importance for

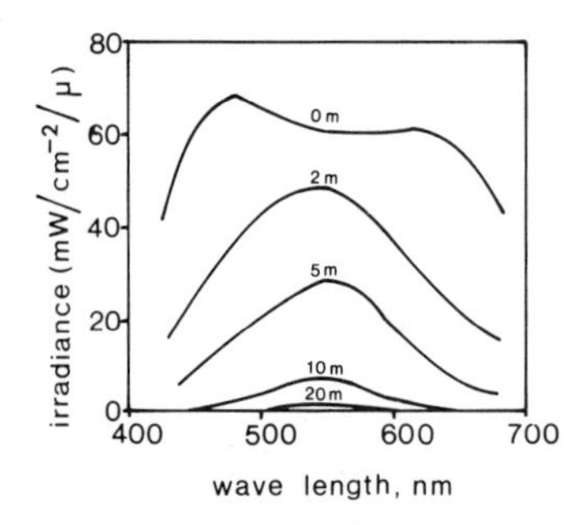

Fig. 19. Spectral distribution of downward irradiance for high solar elevation, Baltic Sea (after Jerlov, 1968).

the taxonomic groups, which will dominate. Fig. 19 shows the spectral distribution of downward irradiance for high solar elevation in the Baltic Sea (according to Ahlquist in Jerlov, 1968). At a depth of 20 m practically only light of between 530 and 570 nm is found. This is in the part of the spectrum where phycoerythrin absorbs. It is thus obvious that red algae especially must be apt to grow in the Baltic at this depth. This is in accordance with the fact that this algal group completely dominates in the lowest part of the photic zone there. On the other hand, brown algae, which are not able optimally to utilize green light, are found at somewhat lesser depths, where the content of fucoxanthin makes them able to utilize light of the wavelength between 470—550, which at these depths is found in considerable quantities — e.g. at a depth of 10 m according to Fig. 19.

In freshwater lakes, where water types similar to the marine coastal water types are often found, we ordinarily do not find a flora of species especially suited to the light quality. Macrobenthic plants are here practically all green. The only "adaptation" to the light quality in the deeper part of the photic zone is shown by the fact that they all seem to have a very high concentration of chlorophyll, just as the marine "deep water" green alga *Chaetomorpha melagonium*. Furthermore, macrobenthic freshwater plants do not seem to penetrate down to quite such low irradiances as the marine species. "Red"-coloured blue-green algae — *Oscillatoria* sp. — belonging to the plankton or the microbenthos, are found considerably deeper in Danish lakes than macrobenthic species. Only very few species of red algae and brown algae are found in freshwater. No real "deep water" macrobenthic flora has developed, although some species prefer the somewhat greater depths.

CHAPTER 7

THE UPTAKE OF CO_2 AND HCO_3^- DURING PHOTOSYNTHESIS

THE CARBON-DIOXIDE SYSTEM

The CO_2 of the air as the C source in the photosynthesis of terrestrial plants has been known since the end of the 18th century, and generally recognized as such from the middle of the 19th century. However, matters have been quite different concerning the C supply of the aquatic plants. This has been due to several circumstances. Until rather recently it was assumed that the aquatic plants assimilated CO_2 from the hydrocarbonate ions by CO_2 being constantly released anew from these ions when the former was removed by the plants.

However, Faurholt (1924) showed that not all processes during the transformation of the carbon-dioxide system are momentary. This applies to the ion processes only, not to the processes of hydration and the corresponding processes of dehydration.

When CO_2 is dissolved in water it is in part hydrated. This takes place in two ways:

$$CO_2 + H_2O \rightleftharpoons H_2CO_3$$

$$CO_2 + OH^- \rightleftharpoons HCO_3^-$$

Only the former process is of real importance at the pH found in sea water. The latter process is only of decisive importance at pH values higher than those found in ordinary seawater.

By far the greater part of H_2CO_3 dissociates according to the process $H_2CO_3 \rightleftharpoons H^+ + HCO_3^-$. A part of the latter may dissociate again according to the process $HCO_3^- \rightleftharpoons H^+ + CO_3^{2-}$. We shall not go into details here concerning the carbon-dioxide system in seawater, but only mention that in ocean water only about 1% of the total CO_2 is found as free CO_2, about 90% as $CO_{2\,HCO_3^-}$ and the rest as $CO_{2\,CO_3^{2-}}$ (cf. Fig. 20). The concentration of H_2CO_3 is so small — about 1/700 of that of the free CO_2 —, that it is normally unnecessary to consider it. The mutual proportion between the various forms of CO_2 in the seawater is dependent on pH, temperature and salinity. In order to study the carbon-dioxide system in seawater in detail we may refer to Buch (1960).

Around the pH of seawater (about 8.2) the hydration and the dehydration process of CO_2 is especially slow. It takes about 80 sec at 18°C to obtain an

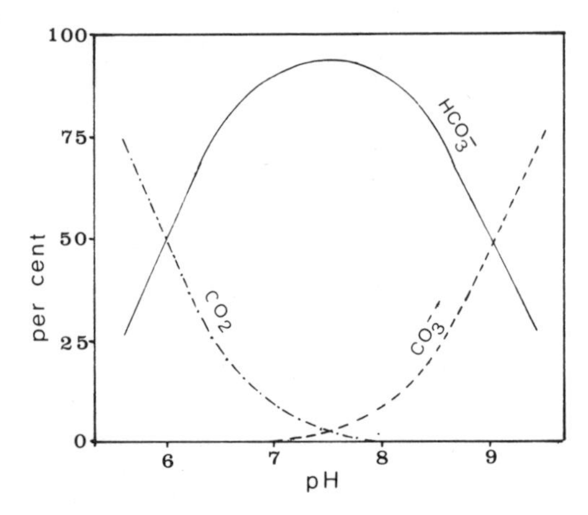

Fig. 20. Relative concentrations of CO_2, HCO_3^- and CO_3^{2-} in seawater of 34‰ salinity at 18°C (after Paasche, 1964).

equilibrium of 90%. In contrast, at the same temperature it takes only 22 sec at pH 6 and 10 sec at pH 10. Such pH values are often found in freshwater but not in the sea.

THE UPTAKE OF THE C-SOURCE BY THE MARINE PLANTS

Raven (1970) has presented an exhaustive review concerning the inorganic carbon source in plant photosynthesis. The special problems concerning water plants were also dealt with by Steemann Nielsen (1960).

If the path for diffusion of CO_2 from the surface of the plant to the chloroplasts is very short, a rapid water exchange around the plant may compensate for the slow release of free CO_2 from hydrocarbonate ions. This is, in fact, the case with the small planktonic algae in which the chloroplasts are found immediately inside the cell wall and the water exchange around the algae is very rapid. This is due to the constant sinking of the algae — about 5 m per 24 hours. This does not at first appear to be a fast sinking rate, but for an alga having e.g. a diameter of 10 μ, it is a very rapid rate of sinking: 50 μ per sec (cf. Chapter 5, phytoplankton).

In larger algae or in the leaves of phanerogames like *Zostera*, the average path for the diffusion from the surface to the chloroplasts is quite long compared with the situation in a small unicellular alga. As the rate of diffusion is inversely proportional to the path of diffusion, not even the most rapid water renewal around the plant can make it possible for free CO_2 in sea water to act as the main C-source. Another source is necessary. This source has turned out to be the bicarbonate ions, which are found at a concentration nearly 100 times as large as that of the free CO_2.

However, ions do not ordinarily diffuse through the plant membranes. An active uptake consuming energy is necessary. In fact, a special mechanism not found in all water plants is necessary for the assimilation of HCO_3^-. Leaves of higher aquatic plants are especially apt to be used when investigating this mechanism. One of the causes for their suitability in such experiments is their ability rapidly to adapt to the various pH values. The rate of photosynthesis can thus be made independent of pH within a wide range. This at least has been the case for the phanerogamic freshwater plants which have been employed for the experiments (cf. Steemann Nielsen, 1947). It is rather doubtful, however, that this would also be the case, e.g., for leaves of the marine *Zostera*. Whereas in many freshwater lakes and streams pH may vary rather much from season to season and even between night and day, the variations are distinctly smaller in marine habitats. Only rather few experiments for measuring the C-sources in marine plants have been made (cf. Table 1 in Raven, 1970).

According to Tseng and Sweeney (1946) a marine filiform red alga seemed to be unable to utilize HCO_3^- as carbon source. Large marine algae like *Laminaria*, *Fucus* and *Macrocystis* must, in accordance with their morphology and anatomy, be forced to utilize hydrocarbonate.

A marine plant which has been shown experimentally to be able to utilize hydrocarbonate ions as the carbon source in photosynthesis is, rather astonishingly, a small plankton alga, the coccolithophorid *Coccolithus huxleyi* (according to Paasche, 1964). This very interesting case will be especially discussed at the end of this chapter. Another claim of HCO_3^--utilization by marine plankton algae (Hood and Park, 1963) has been shown both by Steemann Nielsen (1963) and by Watt and Paasche (1963) to be due to misinterpretation of experimental results.

THE MECHANISM OF HCO_3^--UPTAKE AND UTILIZATION

As no experiments showing HCO_3^- utilization by marine plants can be presented (*Coccolithus huxleyi* is a special case), experiments with the freshwater phanerogamic plant *Myriophyllum spicatum* (according to Steemann Nielsen, 1947) will be discussed instead. Experiments were made both at pH 4.5, where practically only free CO_2 is found, and at pH 9.3, where practically only HCO_3^- is found besides CO_3^{2-} ions, which were shown to have no influence on photosynthesis. At all rates of photosynthesis it was found that variations in pH were without influence on the utilization of free CO_2 and HCO_3^- ions. The utilizations of the two C-sources were rather different, however (see Fig. 21).

Both at a high illumination rate — 33 mW per cm^2 (= about $90 \cdot 10^{15}$ quanta) — where light saturation is found, and at 6 mW, free CO_2 at an optimum concentration is able to give rise to a considerably higher rate of

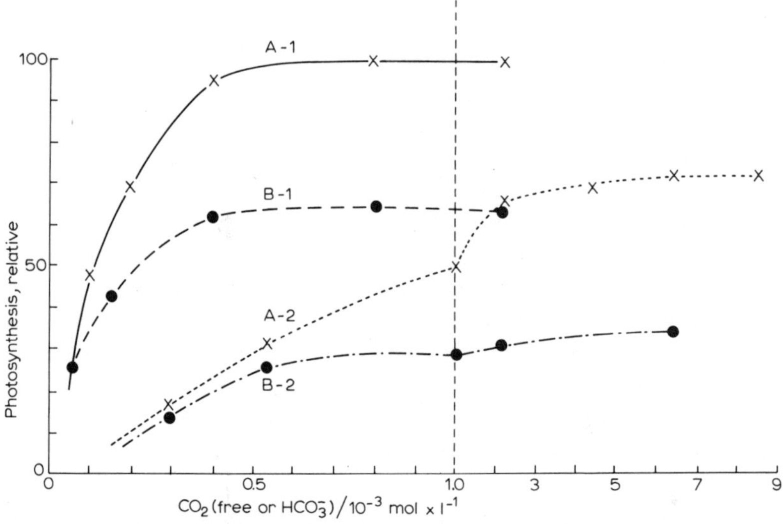

Fig. 21. Photosynthesis in *Myriophyllum spicatum* as a function of the concentration of CO_2 (free or as HCO_3^-). *A*: $78 \cdot 10^{15}$ quanta; *B*: $14 \cdot 10^{15}$ quanta. *1*: pH 4.5; *2*: pH 9.3 (after Steemann Nielsen, 1947).

photosynthesis than CO_2 $_{HCO_3^-}$, when this C-source is also found at an optimum concentration. The explanation is that free CO_2 is able, without use of energy, to diffuse from outside into the plant, whereas HCO_3^-, being an ion, requires energy for the active uptake. ATP produced during photosynthesis is, according to all evidence, the energy donor. It is thus easy to understand that the rate of photosynthesis must be lower when a part of the ATP produced either in the cyclic or the non-cyclic electron transport is drained away to be used for the uptake of HCO_3^-.

By letting a leaf of an aquatic plant function as a partition between the experimental water found partly on the morphologically upper side and partly on the lower side of the leaf, Arens (1936) demonstrated that in light, HCO_3^- and an equivalent amount of Ca^{2+} were absorbed from the lower side and that Ca^{2+} and OH^- were released from the upper side. The experiments were done only qualitatively. They have been repeated on a quantitative basis by Steemann Nielsen (1947). Arens' results were verified in principle. It was shown, however, that a leaf of a higher aquatic plant could assimilate hydrocarbonate not only from the morphologically lower side but also from the upper side. Ca^{2+} and OH^-, on the other hand, were only released from the morphologically upper side. A transport of ions in one direction must thus take place in the leaves in light, when HCO_3^- is the carbon source for photosynthesis.

In the chloroplasts inside the leaf, HCO_3^- is perhaps dehydrated to CO_2 and OH^-. This would be possible due to the presence of the enzyme carbonic anhydrase. Another possibility is that HCO_3^- is directly used for the carboxylation process (see p. 12). In this case OH^- also would be released and must together with Ca^{2+} be transported actively to the upper side of the leaf. In fact, we do not know for sure how the carboxylation process takes place. Several instances indicate that the process perhaps is more complicated than a simple carboxylation by means of either CO_2 or HCO_3^- (cf. Raven, 1970).

The release of Ca^{2+} and OH^- at the upper side is, according to Lowenhaupt (1956), a simple ion exchange between Ca^{2+} from inside and H^+ from outside. As OH^- will constantly be formed from the water outside the plant when H^+ is removed, and as H^+ combines with OH^- inside the plant, the result is the same as if Ca^{2+} and OH^- were excreted.

In algae we have no morphologically upper and lower sides and thus very likely no transport of ions in one direction in the species which, due to their thick thalli, are forced to use HCO_3^- as the carbon source in photosynthesis. This transport in one direction should not be necessary, however. Steemann Nielsen (1947) showed that leaves of higher aquatic plants were able to utilize HCO_3^- as the carbon source, even if the ion could only be taken up at the morphologically upper side.

In very eutrophic, alkaline freshwater lakes the pH during summer may increase up to about 10. Under such conditions the concentration of free CO_2 is far too low even for photosynthesis in small unicellular algae; Österlind (1949) could thus also show that *Scenedesmus quadricauda*, a green alga found in such freshwater lakes, is able to utilize HCO_3^- as a carbon source in photosynthesis. In the sea we never find really high pH values, if we exclude small, very eutrophic rockpools. The ability of planktonic algae found here to use HCO_3^- has never been investigated.

It is remarkable, however, that the coccolithophorid *Coccolithus huxleyi* has been shown by Paasche (1964) to be unable to utilize the free CO_2 found in seawater sufficiently for photosynthesis. HCO_3^- is the main C-source for photosynthesis in this alga. It seems to be rather likely that perhaps all coccolithophorids behave as does *Coccolithus huxleyi* with respect to the utilization of HCO_3^- as the main carbon source. Coccolithophorids may be an old and very special physiological group of algae lacking a mechanism, which in other taxonomic groups of algae provides for the concentration of CO_2 before the proper carboxylation process in the chloroplasts.

SALINITY AND THE UPTAKE OF CARBON

Montfort (1931) investigated the influence of salinity on the rate of photosynthesis by diluting seawater with freshwater. Unfortunately, he did not take into consideration that he altered the carbon-dioxide system at the same time.

Hammer (1968) showed that we have to consider two influences on photosynthesis when the salinity is changed: (a) the indirect influence due to the changed CO_2 system, and (b) the more direct influence due to the changed osmotic pressure.

Steemann Nielsen (1954) explained why certain freshwater species of higher plants were not found in Finland in freshwater but only in the brackish water at the coasts. Whereas the freshwater was practically without any content of HCO_3^-, the concentration of it was sufficiently high in the brackish water to give rise to an adequate rate of photosynthesis.

CHAPTER 8

THE INTERACTION OF PHOTOSYNTHESIS WITH THE OTHER PROCESSES TAKING PLACE IN PLANTS — WITH SPECIAL REFERENCE TO RESPIRATION

INTRODUCTION

If an organism is to be able to survive in the long run in nature, its growth rate must be able to compete with that of the other organisms living in the same ecological niche. If the growth rate under a special set of ecological conditions is to be optimal for an organism, all the processes taking place in it must match each other. In a unicellular alga thus, the rates of photosynthesis, respiration and uptake of P and N must be well matched.

It is well known that, if the N-uptake is too low, many planktonic algae produce fat or carbonhydrates in large quantities instead of primarily producing proteins by means of the assimilated carbon. Under such conditions, growth more or less stops. A population of a species of planktonic algae is normally only able to persist if it is growing optimally. This is due to the grazing by the zooplankton. If another species is better apt to assimilate N for example, it will oust the first species when the concentration of N is very low.

Generally — but of course with many exceptions — the production of new algae is practically the same as their loss due to grazing. In habitats with more or less stable conditions for production, the size of the population of algae is rather constant. This was first shown for the tropical areas of the South Atlantic by the German "Meteor" expedition nearly 50 years ago (Hentschel, 1933—1936).

RESPIRATION IN THE DARK

The rates of photosynthesis and respiration must, as mentioned above, match each other. Although organic matter is only produced due to photosynthesis, the biological functions of both processes are nevertheless closely correlated. They both produce high-energy phosphate, hydrogen donors and pools of intermediates, which can be used for synthesizing more complex molecules. Photosynthesizing cells generally have a relatively lower rate of respiration than cells without the ability to photosynthesize. The number of mitochondria, the organells, in which the final part of respiration in the dark at least takes place, are much more numerous in the latter cells.

Winokur (1948) cultivated eight *Chlorella* strains at three different intensities of illumination and measured both the rates of photosynthesis and respiration.

The results from three of Winokur's strains (chosen at random) are presented in Table III. The rates of photosynthesis and respiration of the algae grown at the various irradiances are given in percentage of the light-saturated rate of photosynthesis in the individual species grown at 6 klux. The rate of respiration in percentage of the light-saturated rate of photosynthesis of the algal species grown at the irradiance in question is shown in brackets. It is interesting to note that the latter percentages only vary little for the same species.

TABLE III

Rate of photosynthesis and respiration at different irradiances (after Winoker, 1948)

	6 klux		2 klux		0.7 klux	
	P.	R.	P.	R.	P.	R.
Chlorella vulgaris var. *viridis*	100	5.8	57	2.8 (4.9)	28	2.0 (7.1)
C. vulgaris (Columbia)	100	10.1	54	5.5 (10.1)	26	2.6 (10.2)
C. luteoviridis	100	8.8	57	5.6 (9.8)	26	2.2 (9.1)

P: photosynthesis; R: respiration.

The same results as appear in Table III have been found in other laboratories working with other species of unicellular algae. All species are able, under ordinary conditions, to adjust the rate of respiration in such a way that it matches the rate of photosynthesis. In Chapters 10 and 11 we shall discuss this matter again and show that an adjustment to a higher rate of saturated photosynthesis, which means a higher concentration of the enzyme in photosynthesis, normally causes a rise in the concentration of all enzymes. Otherwise we could hardly expect the considerable rise in the concentration of protein per cell (cf. p. 87).

PHOTORESPIRATION

Until quite recently, the usual way to measure the rate of respiration in light was to measure the rate immediately after extinguishing the light. Certain experiments by Brown (1953), using the isotope ^{18}O as tracer, had further been explained as showing no influence of light on respiration.

However, current evidence suggests that substantial changes in respiratory processes occur upon irradiation and that under certain conditions, the light respiratory rate may be a significant fraction of the photosynthetic rate. Jackson and Volk (1970) have presented an interesting review of the problem, which until now does not seem to have found a final solution. Therefore, we shall only present some of the seemingly most relevant findings. Already a weak irradiance may restrict dark respiratory processes and induce a light-dependent CO_2 release process, involving, amongst other things, glycolate metabolism through the glycolate oxidase pathway (cf. Chapter 2).

Whereas dark respiration is already saturated at an O_2 concentration of 2%, photorespiration continues to increase with O_2 concentrations of up to 100%. According to Jackson and Volk, there is strong evidence for a progressive increase in photorespiration with increasing irradiance. This very likely indicates that the relationship between photosynthesis and photorespiration remains more or less constant, independent of irradiance. It is likely that the internal O_2 concentration becomes high at a high rate of photosynthesis and that this high O_2 concentration again has the effect that the rate of photorespiration increases.

It has been usual to distinguish between apparent and real rates of photosynthesis, the latter being the observed rate — either by means of O_2 production or CO_2 uptake — plus the rate of respiration measured in the dark. In the light of current evidence concerning photorespiration, it is now not so obvious to use the two terms rigorously. It would be impractical to abandon them completely, however. We must at least maintain the term compensation point, which is the irradiance at which the curve showing the rate of apparent photosynthesis as a function of irradiance intersects the abscissa.

We must further admit that the problems concerning photorespiration are at present by no means solved. Therefore, we have to be cautious in being too definite when discussing the relationship between photosynthesis and respiration. The same is also the case when comparing the rates of other processes, such as e.g. the uptake of other nutrients besides CO_2, with the rate of photosynthesis. We still have much to learn.

In studies of primary production it is normal to use the term "gross production" instead of real photosynthesis and "net production" instead of apparent photosynthesis. When measuring the rate of respiration of the phytoplankton in nature using the oxygen technique (see p. 93) we must further take into account that we include the respiration of bacteria and small zooplanktonic organisms. This kind of respiration is also involved in our measurements of the gross production.

As we have seen, the respiration in the light and in the dark is different. However, dark respiration is essential for all plants living in a climate with day and night. During the latter period, cell division often takes place. If planktonic algae, due to lack of stabilization of the water masses, have to circulate daily down to considerably greater depths than the lower boundary of the photic zone, the rate of dark respiration must be low. This has also been shown by C. Hunding (personal communication, 1974) for a Danish lake during spring. The rate of respiration of the planktonic algae (diatoms) was only 5% compared with that of light-saturated photosynthesis. Ordinarily this percentage is found between 8 and 10 (cf. also Chapter 13).

EXTRACELLULAR PRODUCTS DURING PHOTOSYNTHESIS

Soluble organic products are able to escape from all kinds of water plants during photosynthesis. (Planktonic algae have been investigated by Fogg et al., 1965.) The loss should be up to 35% in oligotrophic waters. Unfortunately, the absolute amounts of extracellular products often seem to be questionable, as several artifacts may influence the results. The usual way of investigating extracellular products is to incubate phytoplankton with ^{14}C bicarbonate of high activity in light and to collect the filtrate after a sample is passed through an ordinary membrane filter. After acidifying the filtrate, this is rapidly bubbled with air for 30 min to remove inorganic ^{14}C. Whereas Fogg et al. (1965) and Steemann Nielsen and Wium-Andersen (1971) dried aliquots of the filtrate, others have placed the bubbled filtrate directly in a scintillation fluor. In marine work the drying of the filtrate is impractical due to the large amount of salt present. However, Sharp and Renger (1973) have shown that over 0.02% of added inorganic ^{14}C can remain in a seawater sample even after bubbling for 30 min. This background, which is mistaken for organic ^{14}C in the filtrate, will account for a high apparent excretion in oligotrophic oceanic water, as e.g. in the Sargasso Sea.

Another way of artificially causing excretion of organic matter is by introducing small amounts of poison into the experimental water. The original description of the preparation of ^{14}C ampoules stressed the use of glass-distilled water and the production of $^{14}CO_2$ by distillation. This procedure has always been followed by the International Carbon-14 Agency. However, the frequently used manual of Strickland (1960) amongst others recommends a simple dilution of commercially bought solutions of $NaHCO_3$ with ordinary distilled water. An average of 260 μg of Cu per liter was found in 350 samples of distilled water. If such average distilled water is used when manufacturing ^{14}C ampoules, and a 1-ml ampoule is added to 100 ml experimental water, the amount of Cu added will be about 3 μg/l. Furthermore, other toxic substances may be introduced into the ampoules by commercially bought $NaHCO_3$. According to Steemann Nielsen and Wium-Andersen (1971), diatoms (but not *Chlorella*) excrete organic matter in the presence of copper. It should furthermore be added that the release of organic matter can be high when photosynthesis is inhibited by high irradiance.

Finally, it must be added that organic matter may be brought into the filtrate secondarily, if the filters are rinsed for inorganic C — found e.g. in coccoliths — by sucking dilute HCl through the filters instead of treating them with fuming HCl (see e.g. Strickland, 1960).

CHAPTER 9

THE RATE OF PHOTOSYNTHESIS AS A FUNCTION OF IRRADIANCE

INTRODUCTION

As expounded in Chapter 2, two kinds of process are taking part in photosynthesis, i.e., photochemical processes and enzymatical processes. The rates of the photochemical processes depend on the light absorption by the photosynthetic pigments, and therefore on both the concentration of the pigments (at least in unicellular algae) and on the strength of irradiance. On the other hand, the rates of the enzymatic processes depend on the concentration of active enzymes and on the temperature. The rate of the overall process may be limited by the rates of either the photochemical or the enzymatic process. Perhaps the most important consequence of adaptation — as will be shown in the next chapter — is the possibility of matching to some extent the photochemical and the enzymatical processes under the prevailing ecological conditions.

PHOTOSYNTHESIS CURVES

The slope of the initial part of the curve showing the rate of photosynthesis as a function of irradiance (see e.g. Fig. 22) is a function of the photochemical part of the photosynthesis. On the other hand, the horizontal part represents the maximum rate of the enzymatic processes at the prevailing temperature. Therefore, the irradiance at which the initial slope and the horizontal part of the irradiance—photosynthesis curve intersect (cf. Fig. 22) describes, to a certain degree, the ratio between the two kinds of process. This quantity, introduced as I_k by Talling (1957), is as will be shown in the next chapter, an important means of describing the physiological adjustment of an algal population.

In laboratory work with unicellular algae it is easy to present the rate of photosynthesis per definite units, such as number of cells, dry weight, or weight of chlorophyll. In Fig. 23 the rate of photosynthesis is given as a function of irradiance in *Chlorella vulgaris* at 20° C, grown at either 7.5 or 75 $\cdot 10^{15}$ quanta, the number of cells being the unit in Fig. 23A, the weight of chlorophyll in B and the dry weight in C. It is obvious that the curves vary *inter se* according to the unit applied.

Another way to present irradiance—photosynthesis curves is to normalize

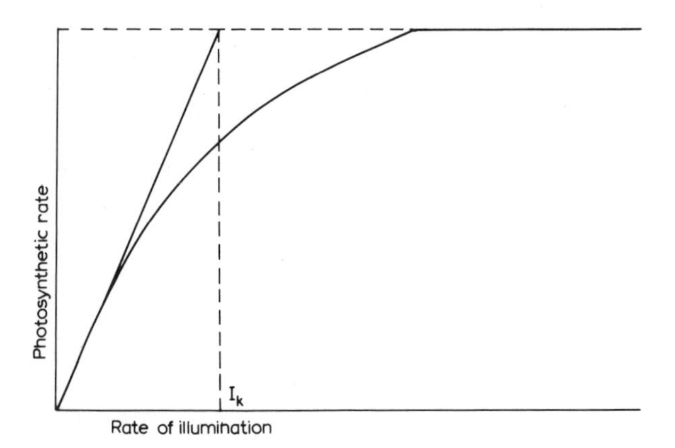

Fig. 22. Photosynthesis as a function of irradiance showing the position of I_k (after Talling, 1957).

each curve with respect to the maximum rate of photosynthesis (cf. Fig. 37). When working with plankton collected in nature, it is difficult in many cases to obtain reliable units to which the rate of photosynthesis can be related.

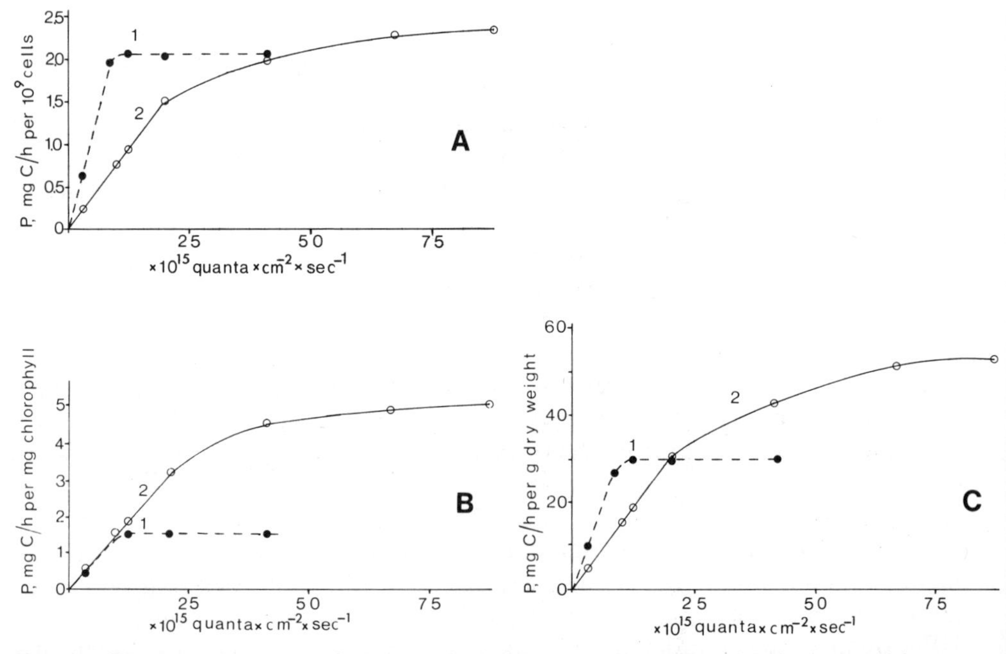

Fig. 23. Photosynthesis as a function of irradiance in *Chlorella vulgaris* grown either at $7.5 \cdot 10^{15}$ quanta (1) or at $75 \cdot 10^{15}$ (2). In A the rate of photosynthesis is given per number of cells, in B per mg chlorophyll and in C per g dry weight (after Steemann Nielsen, 1962).

Numbers of cells are unrealistic if the plankton, as is usual, consists of many different species of varying sizes. Chlorophyll, on the other hand, can often be used with success (cf., e.g., Ichimura et al., 1962).

Since photosynthesis is a photochemical process, the proper unit to use for irradiance is quanta per second and per surface unit (cf. Chapter 4). This is particularly important when working in nature. In the lower part of the photic zone in coastal water, we may only find green light and in oceanic water only blue light. When working in the laboratory with various light sources, for some algae it is of little importance whether the rate of photosynthesis is given per units of quanta or per units of energy (cf. Steemann Nielsen and Willemoës, 1971). In Figs. 24 A and B curves are presented for the green alga *Chlorella pyrenoidosa*. The rates of photosynthesis as a function of irradiance are rather alike in fluorescent light (Philips W/33) and in incandescent light, whether the intensity is measured in quanta or in energy. In the blue-green alga *Coelosphaerium* sp. (cf. Fig. 25), on the other hand, an obvious difference occurs, even if the illumination is measured in quanta.

In Fig. 26 curves are presented for the rate of photosynthesis of *Nitzschia palea* as a function of irradiance: (a) in blue light — as in the lowest part of the photic zone in the clearest ocean water; and (b) in green light — as in the

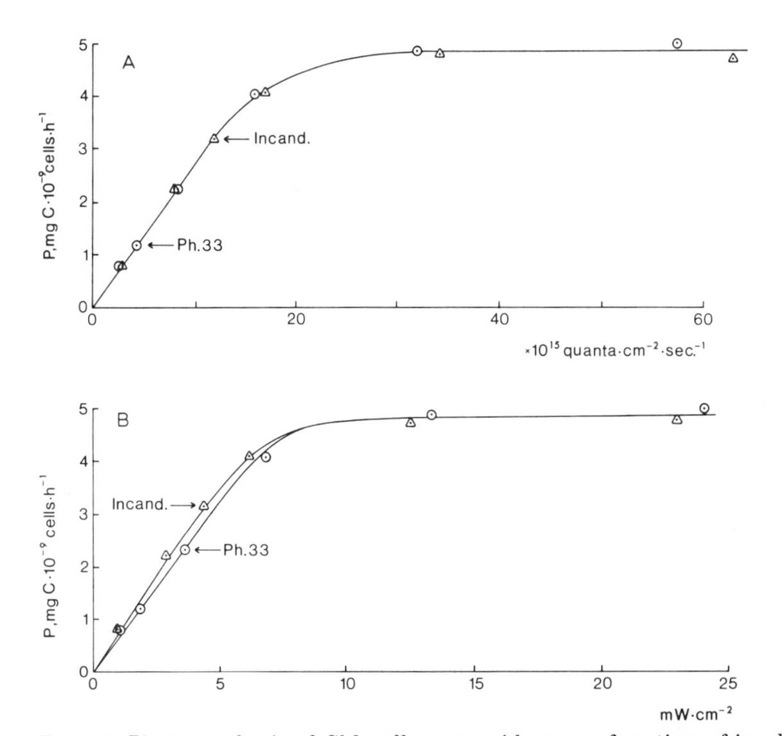

Fig. 24. Photosynthesis of *Chlorella pyrenoidosa* as a function of irradiance (incandescent or fluorescent light, Philips 33), 20°C. In A the irradiance is measured in quanta, in B in energy (after Steemann Nielsen and Willemoës, 1971).

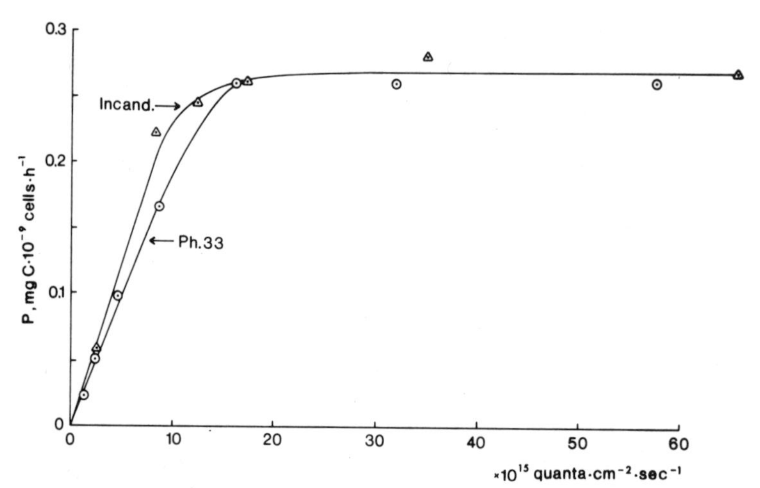

Fig. 25. Photosynthesis of *Coelosphaerium* sp. as a function of incandescent or fluorescent light (Philips 33) measured in quanta, 20°C (after Steemann Nielsen and Willemoës, 1971).

lowest part of the photic zone in coastal water, optical water type 3. Irradiance was measured in quanta. As standard of reference photosynthesis in fluorescent light (Philips W/33) was used and this corresponds rather well

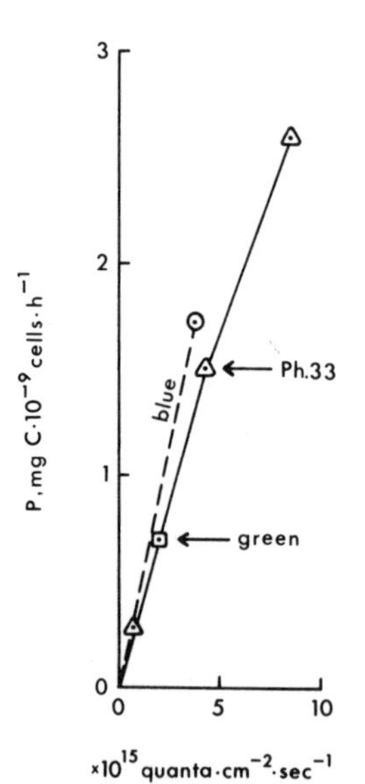

Fig. 26. Photosynthesis of *Nitzschia palea* as a function of irradiance measured in quanta, 20°C. Using a blue filter, the spectrum was nearly the same as at 100 m in ocean water, optical type I. By means of a green filter, the spectrum was nearly the same as at a depth of 20 m in coastal water, optical type 3. Philips 33 produced light, in spectral composition rather near to that found at 8 m in coastal water, optical type 7 (after Steemann Nielsen and Willemoës, 1971).

with the light in the lower part of the photic zone in coastal water, optical water type 7. The diatom *Nitzschia palea* was used for these experiments because diatoms are able to absorb the quanta more or less equally well over the whole spectral range of importance for photosynthesis. Fig. 26, thus, shows that in diatoms the rate of photosynthesis per number of quanta is more or less independent of the optical water type, even in the lowest part of the photic zone.

CHLOROPHYLL CONCENTRATION AND RATE OF PHOTOSYNTHESIS AT LOW IRRADIANCES

Gabrielsen (1948) has especially stressed the importance of studying the influence of chlorophyll concentration on the rate of photosynthesis at a weak irradiance, where the rate of the overall photosynthesis process is limited by the rate of the photochemical part of the processes. As shown by him, chlorophyll is normally present in considerable excess in leaves of higher plants. The chlorophyll content per surface unit of leaves in terrestrial plants is, according to Seybold and Weissweiler (1942), higher in "sun" leaves than in "shade" leaves. Investigations concerning the influence of chlorophyll concentration can only be made using mutations with leaves extremely poor in chlorophyll. Dilute suspensions of planktonic algae are still better suited for such investigations, as the pigments here are never found in real excess (cf. Steemann Nielsen, 1961).

A considerable number of planktonic green algae and diatoms have been investigated in the author's laboratory. At $2.5 \cdot 10^{15}$ quanta incandescent light they have given, per mg chlorophyll *a*, a rate of photosynthesis of about 0.4—0.5 mg C/h, when cultivated under ordinary conditions.

Some exceptions to the rule have been found, however. *Chlorella vulgaris*, grown at a weak irradiance ($7.5 \cdot 10^{15}$ quanta) but provisionally transferred to a strong irradiance for a few hours, suffers inactivation of a part of the photochemical mechanism, whereby the rate of photosynthesis at $2.5 \cdot 10^{15}$ quanta decreases considerably (cf. Steemann Nielsen, 1962, and p. 75).

Another exception to the relatively constant rate of photosynthesis per mg chlorophyll *a* at $2.5 \cdot 10^{15}$ quanta is found in extremely nutrient-deficient cultures. In *Skeletonema costatum*, with which Steemann Nielsen and Jørgensen (1968a) have obtained in normal experiments an average rate of fully 0.50 mg C per mg chlorophyll *a* at $2.5 \cdot 10^{15}$ quanta, the rate was 0.19 in a culture pronouncedly deficient in phosphorus, and 0.09 in a culture pronouncedly deficient in nitrogen.

Finally, poisons may lower the rate per mg chlorophyll *a* at both high and low light intensity as shown, e.g., by Weller and Franck (1941); compare Fig. 27, where Cu is used as the poison.

A content of pigments such as fucoxanthin will increase the rate in dia-

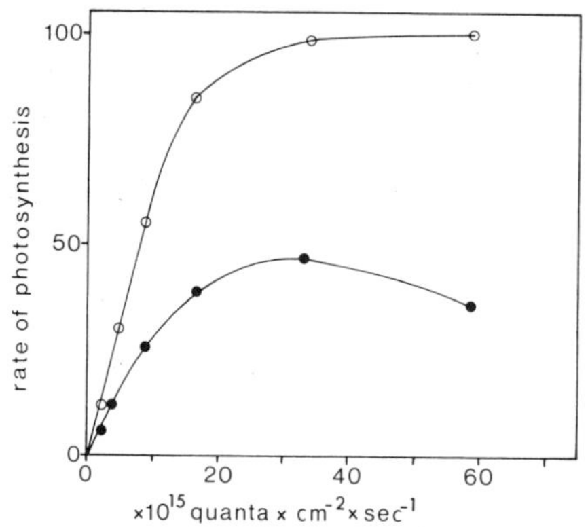

Fig. 27. Photosynthesis as a function of irradiance, 20°C. ○: without addition of Cu; ●: addition of 3 μg Cu/l (after Steemann Nielsen and Wium-Andersen, 1971).

toms. It is impossible for algae such as diatoms to give a precise value for the rate of photosynthesis per mg chlorophyll *a* at $2.5 \cdot 10^{15}$ quanta in incandescent light. The variation of the value given is due not only to the experimental scatter, but also to the variation in the concentration of the other pigments. Finally, the content of chlorophyll *a* may vary rather considerably due to the techniques used for measuring it.

Some variations in the value must therefore be expected. Large deviations from the value of 0.4 mg C per mg chlorophyll *a* and per hour, however, are indications of either faults in the experiments or influences of, e.g., nutrient-deficient or poisoned cultures (cf. above).

The ratio between the light-saturated rate of photosynthesis and the weight of chlorophyll has been used by several scientists working with plankton algae — e.g. Gessner (1944). However, this ratio is very variable even if all chlorophyll is active and it illustrates, in fact, the physiological adjustment of the photosynthetic process — the ratio between the photochemical and the enzymatic processes in photosynthesis. It shows, thus, more or less the same as I_k (cf. p. 53).

LIGHT SATURATION AND DECREASE OF PHOTOSYNTHESIS AT STRONG IRRADIANCE

The rate of photosynthesis will always stop increasing at a certain irradiance when the maximum rates of the enzymatic processes have been achieved. We then have light saturation. However, if we continue to increase irradiance, sooner or later the rate of photosynthesis will start to decrease.

It is a well-known fact that most photo-autotrophic plant species are seriously affected by stronger irradiances than those normally found in their habitats (cf. Stålfelt, 1960). Species of higher plants normally seem only to

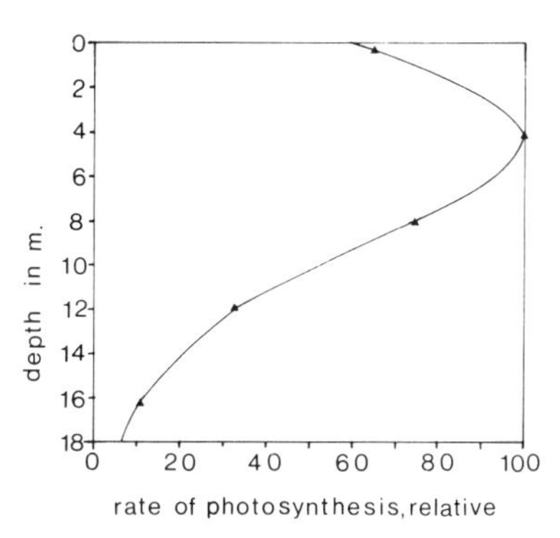

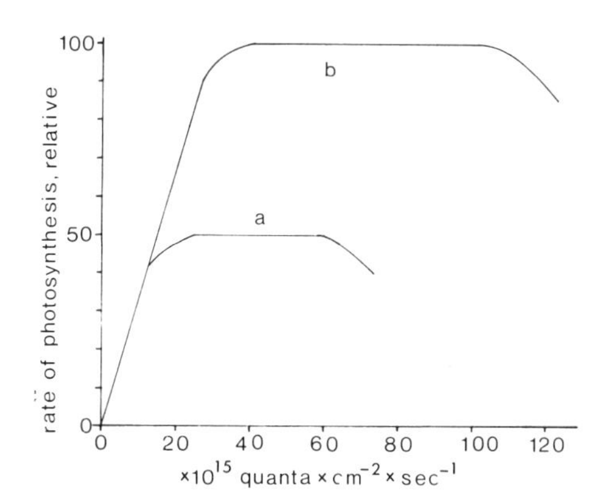

Fig. 28. The relative rate of photosynthesis in water collected at a depth of 4 m and exposed at different depths, coast of South Africa (after Steemann Nielsen and Jensen, 1957).

Fig. 29. Photosynthesis per unit of chlorophyll as a function of irradiance in *Chlorella vulgaris*, grown either at 7.5 (*a*) or at 75 · 10^{15} quanta (*b*).

be influenced under such conditions. However, planktonic algae normally behave differently. The rate of photosynthesis on bright days is depressed near the surface. If bottles containing ordinary freshly collected surface water are suspended at different depths within the photic zone on a bright day, the maximum rate is found at the depth where about 30—50% of the irradiance at the very surface is found (cf. Fig. 28). This is the case both in the tropics and at higher latitudes during summer. If the water from the lower part of the photic zone is used instead for such experiments, the maximum rate ordinarily is found at still greater depths.

In laboratory experiments, above a certain irradiance, a decrease in photosynthesis is also observed. If the algae have grown at a really strong irradiance, the decrease in the photosynthesis rate first takes place at a relatively strong irradiance, whereas it will be found at a relatively low level, if the algae have grown at a weak irradiance (Fig. 29). Concerning the influence of the ultraviolet part, see p. 63.

If the growth conditions for the algae have been somewhat adverse for some reason, the decrease in photosynthetic rate is much more pronounced. Fig. 30 shows an irradiance—photosynthesis curve for *Chlorella vulgaris* grown at 7.5 · 10^{15} quanta in an N-deficient medium. It is obvious that the decrease at stronger irradiances is much more pronounced than in the corresponding experiments with normal *Chlorella* cells shown in Fig. 29. In Fig. 27 an experimental series was presented showing that poison — a small concentration of Cu — has the effect that the rate of photosynthesis is decreased at a definitely weaker irradiance than under ordinary conditions.

In the next chapter we shall discuss how the decrease in the rate of photosynthesis is effected.

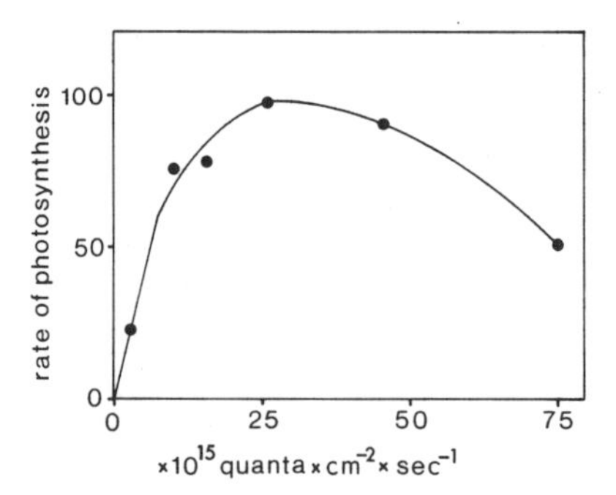

Fig. 30. Photosynthesis (relative) as a function of irradiance in *Chlorella vulgaris* grown at $7.5 \cdot 10^{15}$ quanta in an N-deficient medium, 20°C (after Steemann Nielsen, 1962).

DIURNAL VARIATIONS IN PHOTOSYNTHETIC ACTIVITY

Since the work of Doty and Oguri (1957) and Yentsch and Ryther (1957) we know that a diurnal variation in photosynthetic activity and chlorophyll content of phytoplankton may take place in nature, especially at low latitudes and near the surface, but also to some extent at higher latitudes. Ordinarily the phenomenon is described as the "afternoon depression".

Several factors contribute to these fluctuations; an internal rhythm in the algae is one of the causes. In most cases planktonic algae in nature appear to be synchronized to some extent. Sorokin (1957) was the first to show the existence of considerable variations in the irradiance—photosynthesis curve during the different periods (growth stages) of synchronous cultures of *Chlorella*. A shift between 9 hours in light and 15 hours in the dark was necessary for producing a synchronous culture. Pirson and Lorenzen (1966) have presented a review concerning synchronizing of algae.

By using a partly synchronized culture of the marine diatom *Skeletonema costatum* — 12 hours of light and 12 hours of darkness — Jørgensen (1966) could show the same phenomenon. The rate of light-saturated photosynthesis per mg chlorophyll *a* was about twice as high during the middle of the light period as during the middle of the dark period.

A proper synchronous culture, where the division of the single cells in the culture takes place at practically the same time, seems by no means to be necessary in order to give rise to variations in the photosynthetic rate, between a light and a dark period. Fig. 31 shows, according to Steemann Nielsen and Wium-Andersen (1972), the increase in cell number of a culture of *Nitzschia palea* growing in a programmed light ($10 \cdot 10^{15}$ quanta) and darkness regime of 12 hours light and 12 hours darkness. The diatom started to divide 2 hours after the start of the light period and ceased about 3 hours before the end of the dark period. During these 19 hours the division rate was rather constant. Fig. 32 presents a series of experiments where the rate

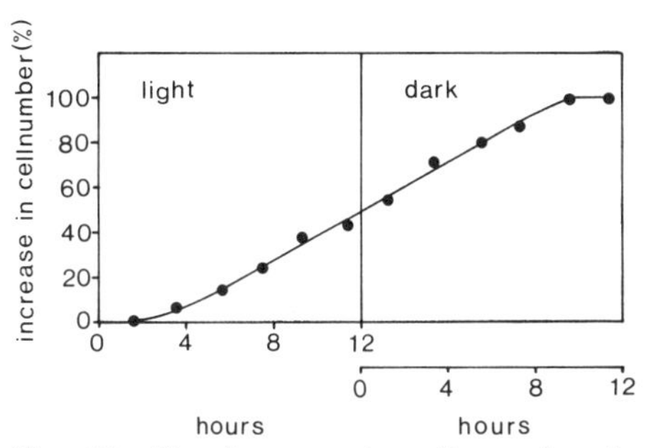

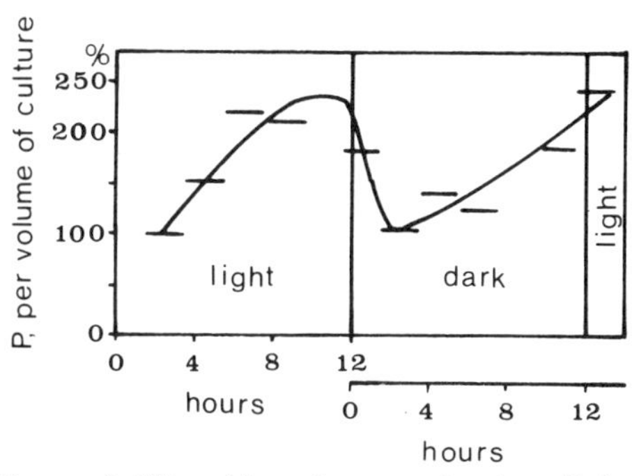

Fig. 31. The increase in cell number in a culture of *Nitzschia palea* growing in a light ($10 \cdot 10^{15}$ quanta) and dark regimen of 12 hours each, 20°C (after Steemann Nielsen and Wium-Andersen, 1972).

Fig. 32. The rates of light-saturated photosynthesis (experiments lasting 2 hours) per equal volume of a culture of *Nitzschia palea* growing in a programmed light and dark regimen of 12 hours light ($10 \cdot 10^{15}$ quanta) and 12 hours dark (after Steemann Nielsen and Wium-Andersen, 1972).

of photosynthesis was followed during the same 24 hours at light saturation. The same volume of water from the culture was added to the experimental water (for measuring photosynthesis) during the whole series. The rate of photosynthesis increased during the whole light period, but immediately after the beginning of the dark period it decreased, showing an absolute minimum about 2—4 hours into the dark period. The rate then increased again, reaching at the transition to the new light period about the same value as that found during the last hours of the first light period. *Skeletonema*, according to Jørgensen (1966), did not show any decrease in the absolute rate of photosynthesis during the dark period, although I_k decreased. The difference in behaviour between the two species of diatoms is very likely in some way interrelated with the fact that *Nitzschia* grows very badly in continuous light at 20° C, whereas *Skeletonema* grows well at 20° C under such conditions. The light—dark regime must effect a mechanism preventing the photosynthetic system from functioning optimally during most of the dark period in *Nitzschia palea*. It seems rather unlikely that the division process has anything to do with the reduction of photosynthesis.

These experiments give a warning not to start primary production experiments during late afternoon or at night.

The fluctuations in I_k in *Skeletonema costatum*, such as shown by Steemann Nielsen and Jørgensen (1968b) (see Table IV) were due to the fact that during growth of the diatom the period for the production of chlorophyll and the period for the production of photosynthesis enzymes — measured by means of the rate of light-saturated photosynthesis — were displaced mutually.

Despite the fluctuations, the difference in nature in I_k between surface

TABLE IV

I_k of partly synchronized *Skeletonema costatum* grown 12 hours with $7.5 \cdot 10^{15}$ quanta and 12 hours in the dark (20° C)

Duration of the photosynthesis experiments (number of hours after the beginning of the light period)		$I_k \cdot 10^{15}$ quanta
light period	3	36
	$4\frac{1}{2}$	33
	5	33
	$11\frac{1}{2}$	18
dark period	13	18
	$16\frac{1}{2}$	18
	17	18
	$23\frac{1}{2}$	23

plankton and plankton from the lower part of the photic zone is always pronounced, except for the winter at higher latitudes and areas without vertical stabilization of the water masses. In the next chapter we shall discuss the variation of I_k due to the adaptation of the planktonic algae at the various depths. In the Danish waters during the height of summer, we may tentatively state that I_k at the surface and in the lower part of the photic zone varies within the ranges of 20—40 and $5—10 \cdot 10^{15}$ quanta, respectively.

THE INFLUENCE OF DAY LENGTH

Several authors (e.g., Pirson, 1957; Tamiya, 1957; Sorokin and Kraus, 1959) have shown inhibition of growth under long day lengths. This day length "oversaturation" may be most pronounced at a strong irradiance. As will be shown in Chapter 11, temperature is an important factor determining the optimum length of the day.

Castenholz (1964) has discussed the ecological consequences of the adjustment to varying day length in unicellular algae (sessile diatoms from the Oregon coast). The growth of *Fragilaria striatula*, e.g., was dependent on the length of the day. Doubling rate was significantly lower during short days than long days both below and above saturating light. *Melosira moniliformis* showed less dependence on day length, but was inhibited by strong irradiances during a long day. *Biddulphia aurita* had essentially the same growth rates during 9- and 15-hour lengths of day at all irradiances used. Each species in nature has a somewhat different pattern of seasonal abundance on the Oregon coast, where the temperature only varies little around 11° C, the temperature used for the experiments.

Fragilaria was found abundantly between April and October. During the intervening period it disappeared almost completely. *Melosira* and *Biddulphia*, on the other hand, were common throughout the year. In summer, however, the first species definitely occurred in shaded locations. This is in good agreement with the growth experiments.

In Chapter 11 it is shown that the marine planktonic diatom *Skeletonema costatum* is independent of the length of the day at higher temperatures but demands a "short" day at a low temperature.

THE INFLUENCE OF ULTRAVIOLET LIGHT

Ultraviolet light down to wavelengths of about 350 nm reaches the surface of the sea in considerable quantities in clear weather (see Fig. 5). The further penetration down into the water depends on the optical type of the water. Table V shows the percentage of the surface irradiance at a depth of 1 m for various wavelengths in the ultraviolet part of the spectrum, compared with the irradiance immediately below the surface.

At a depth of 10 m in the most transparent ocean water (type I), 22% of the surface irradiance at 310 nm is found. This means that ultraviolet light of this wavelength penetrates relatively deep down in clear ocean water. On the other hand, in all coastal water types, even in the clearest (1), we can fully disregard this light at a depth of only some few meters. The presence of yellow matter in coastal waters (see p. 20) is the cause of the poor penetration of ultraviolet light.

Concerning the influence of ultraviolet light on marine photoautotrophic plants, we may distinguish between the ability to promote photosynthesis and the detrimental effect on the plant, inclusive of its photosynthetic apparatus. Johnson and Levring (1946) showed that species of green, brown and red algae were able to photosynthesize at 366 nm. The rate, however, was low, and it is impossible to make out whether a detrimental effect also

TABLE V
Irradiance transmittance (%/m) for surface water of different water types (after Jerlov, 1968)

Water type	310 nm	350 nm	375 nm
I	86	94	96
III	50	71	79
1	16	32	54
5	3	10	21
9		1.5	4.7

played a role. All the species investigated came from coastal water. No investigations have been published concerning the photosynthesis in ultraviolet light of planktonic algae coming directly from clear oceanic water.

Coastal surface plankton is definitely adversely influenced by the ultraviolet light found near the surface. This was shown by Steemann Nielsen (1964c). Surface water from the bight off the Friday Harbour Biological Laboratories, State of Washington, was sampled on 7th July and placed at 09:30 h in the sun for 90 min in two flat, open containers of glass cooled by running water. On the top of one of the containers a plate of clear glass — 3 mm thick — was placed. Immediately afterwards photosynthesis experiments with the water were made in incandescent light for 2 hours. Curves showing the rate of photosynthesis as a function of irradiance are presented in Fig. 33. They show that the presence or absence of the 3 mm thick glass plate over the seawater illuminated in sunshine is of decisive importance for the succeeding rate of the photosynthesis. Both the rate of light-saturated photosynthesis and the initial slope of the curve are affected. This is an indication that enzymes active in photosynthesis are destroyed and that the photochemical mechanism is also affected by the ultraviolet light (which is absorbed by a glass plate, 3 mm thick).

In other experiments made simultaneously it was shown that plankton from depths to which ultraviolet light never penetrates is sensitive to the more long wave part of the ultraviolet light at the surface, even if it is enclosed in ordinary clear glass bottles. Surface plankton on the other hand, seems to be influenced much less by such ultraviolet light under the same conditions. The effect on plankton from the deeper parts of the photic zone is of importance when the simulated *in situ* technique is used for measuring the rate of primary production (cf. p. 94).

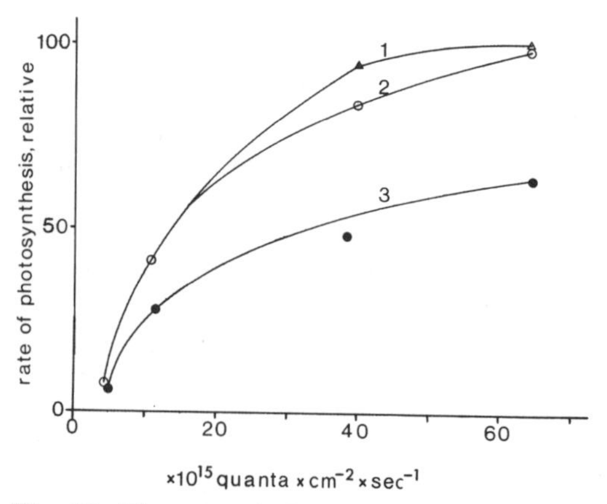

Fig. 33. The rate of photosynthesis (relative) as a function of irradiance. Surface plankton. *1*: after 90 min in the dark; *2*: after 90 min in the sun, but covered with a plate of glass; *3*: as *2*, but without a plate of glass (after Steemann Nielsen, 1964a).

In freshwater plants, both unicellular algae and phanerogams, experiments have given similar results; for a review see Gessner (1955).

When making *in situ* measurements with phytoplankton in glass bottles for measuring the rate of primary production, a distinct inhibition of the rate of photosynthesis near the surface is found in clear weather (see e.g. Fig. 28). However, this can scarcely only be due to the ultraviolet light. An inhibition is also found when using strong irradiances in laboratory experiments with incandescent light, which contains practically no ultraviolet light.

It is obvious from the facts presented concerning the influence of ultraviolet light, that measurements of the rate of photosynthesis at the surface obtained by means of bottles often do not represent the true value in the sea.

Biebl (1952) investigated the influence on macrobenthic algae of very short wave ultraviolet light (310—230 nm). Although this light is of little ecological importance for waterplants, the results are still of interest. The red alga *Porphyra umbilicalis*, which is found in the highest parts of the tidal zone, was the species best adapted to survive the irradiation.

THE EFFECT OF POLARIZED LIGHT

There are contradictory views in the literature as to the part played by polarized light in photosynthesis. However, McLeod (1957) has presented convincing evidence that the net photosynthesis and chlorophyll *a* synthesis of the marine planktonic alga *Dunaliella euchlora* shows a gradation in response to various types of polarized light. Right-circularly polarized light increases photosynthesis and chlorophyll *a* synthesis as compared with light through a neutral-density filter, whereas the effect is reversed with left-circularly polarized light. This suggests that the receptor pigments are anisotropic and exhibit circular dichroism. It cannot be excluded that the phenomenon may have some ecological significance in the photosynthesis of some marine species under natural conditions.

CHAPTER 10

PHYSIOLOGICAL ADAPTATION TO IRRADIANCE

INTRODUCTION

In this chapter we shall discuss the adaptation of algae to different irradiances. Although the different species vary rather much concerning the extent of their adaptation, they are all able to adapt to some degree.

Investigations on this kind of adaptation of the photosynthetic apparatus in aquatic plants were started by Ruttner (1926) using the freshwater higher plant *Elodea*. He was followed by Harder (1933) among others. Truly physiological investigations were started by Gessner (1938), with some freshwater higher plants and by Myers (1946) who used cultures of *Chlorella*.

The word "adaptation" is of course used in the physiological sense, i.e. for a physiological adjustment to the surrounding conditions. Plant physiologists seem to have used the word adaptation exclusively with such a meaning at least since the end of the last century. Biologists working with genetics, on the other hand, understand by adaptation a hereditary alteration adjusting the organisms to the surroundings. For physiologists then adaptation is "phaenotypical" in contrast to the "genotypical" meaning as used in genetics.

The photosynthetical mechanism is, of course, not the only mechanism which adapts according to the ecological conditions. Respiration must especially be mentioned. Respiration and photosynthesis ordinarily adapt concordantly (cf. p. 49). In plants growing harmoniously the two processes must be adjusted in such a way that they match each other.

As mentioned in Chapter 9, the irradiance—photosynthesis curve is an important means of describing the physiological adjustment of a plant. It is, however, very important to keep in mind that it only gives an instantaneous measure of the ability of the plant to react to various experimental conditions of short duration. Most important for a plant is the rate of photosynthesis at the irradiance at which it grows.

THE IMPORTANCE OF ADAPTATION FOR THE ECONOMY OF THE PLANKTONIC ALGAE

From terrestrial ecology we know that shade leaves have the best economy at weak irradiances and sun leaves at strong irradiances (cf., e.g., Boysen

Jensen, 1918). The advantage of the sun leaves in full light lies in the fact that they can utilize strong irradiances due to the high level of their light saturation. The rate of the real photosynthesis of shade leaves per leaf area at low light intensities is more or less the same as that of sun leaves.

With the exception of those extremely poor in chlorophyll, the rate of photosynthesis in leaves, even at weak irradiances, is independent of chlorophyll concentration (cf. Gabrielsen, 1948). When shade leaves have a better economy than sun leaves in habitats with a weak irradiance, this is due to the lower rate of respiration. We may state that the rate of ιespiration is normally proportional to the rate of light-saturated photosynthesis (cf. p. 50). This is not only true for leaves of terrestrial plants but also for planktonic algae (cf. Winokur, 1948). This article was discussed in Chapter 8.

In contrast with the leaves of terrestrial plants, many species of planktonic algae have one more means of increasing the economy at a weak irradiance. They are able in this case to raise the rate of photosynthesis per cell by increasing the quantity of chlorophyll.

As a general rule, we may state that unicellular algae adapted to strong irradiance have increased the maximum rates of the enzymatic processes relatively to the potential rates of the photochemical processes (which in principle are the same as the concentration of the photosynthetical pigments). In *Chlorella pyrenoidosa* and *C. vulgaris* the adaptation is mainly effected by varying the pigment content per cell. Fig. 34 presents the concentration of chlorophyll *a* in mg per 10^9 cells of *C. pyrenoidosa* in cultures grown in continuous incandescent light between $0.5 \cdot 10^{15}$ and $36 \cdot 10^{15}$ quanta (cf. Steemann Nielsen and Jørgensen, 1968a). A variation by a factor of about 10 is observed.

The very low chlorophyll concentration at the relatively moderate irradiance of $36 \cdot 10^{15}$ quanta is not due to a temperature of 20°C, which is relatively low for the present species. At 29°C and the same irradiance, 0.30

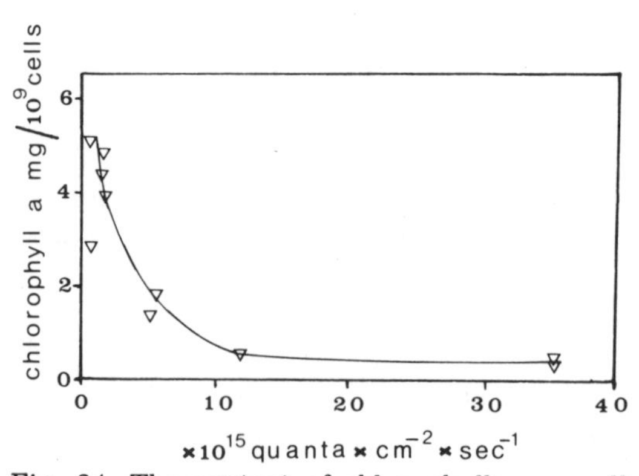

Fig. 34. The content of chlorophyll *a* per cell in *Chlorella pyrenoidosa* (211/8b) as a function of the irradiance at which the algae were grown. Continuous incandescent light, 20°C (after Steemann Nielsen and Jørgensen, 1968a).

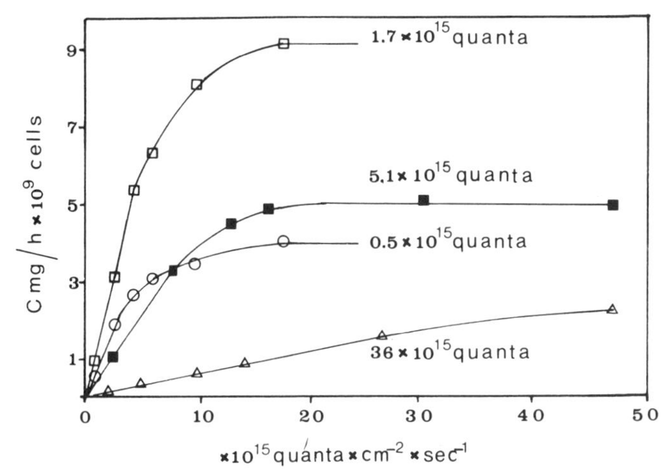

Fig. 35. The rate of photosynthesis per cell as a function of irradiance in *Chlorella pyrenoidosa* (211/8b) grown in continuous incandescent light (different irradiance), 20°C (after Steemann Nielsen and Jørgensen, 1968a).

mg chlorophyll *a* per 10^9 cells was found. This is practically the same as at 20° C. Neither is the quality of light important. Experiments using other light sources have given the same results. The result may, however, be partly due to the use of continuous light.

In Fig. 35 irradiance—photosynthesis curves — per number of cells — are given for *Chlorella pyrenoidosa* grown in continuous light of 36, 5.1, 1.7 and $0.5 \cdot 10^{15}$ quanta. They show that adaptation has taken place at all irradiances. At the two weakest irradiances used for growing *Chlorella pyrenoidosa*, the cells are growing abnormally large before dividing. They produce 16—32 autospores in contrast to the normal 4—8. The very high chlorophyll concentration per cell (cf. Fig. 34) is at least partly due to this circumstance. The very high rate of light-saturated photosynthesis per number of cells in the algae grown at $1.7 \cdot 10^{15}$ quanta is probably also a consequence of the large size of the cells. In the algae grown at the still weaker irradiance of $0.5 \cdot 10^{15}$ quanta, the same result was not observed. In this case, the alga is near to the weakest irradiance under which it can grow at all.

Above we have only described one of the two types of adaptation to different irradiances: the *Chlorella* type, which seems to be found mostly among the green algae. As shown above, it adapts to a new irradiance mainly by changing its pigment content. Jørgensen (1964, 1969) has described another type: *Cyclotella*. This is a diatom and the type has until now mostly been found among diatoms. This type adapts by changing the light-saturated rate, which means that the content of enzymes active in photosynthesis changes; the light-saturated rate is much higher when the algae have been grown at a strong irradiance than when they have been grown at a weak one. The two adaptation types are not sharply separated. Transition types occur.

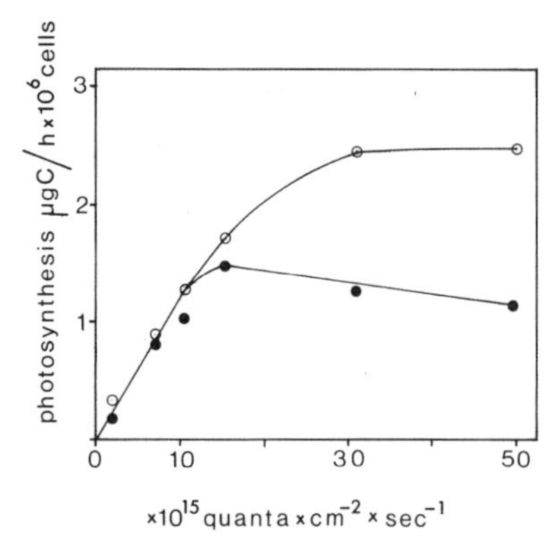

Fig. 36. Photosynthesis of *Nitzschia palea* as a function of irradiance, 20°C. ●: grown at $5 \cdot 10^{15}$ quanta; ○: grown at $50 \cdot 10^{15}$ quanta (after Jørgensen, 1969).

Jørgensen (1969) has presented lists of algal species belonging to the various types. Figs. 23 and 36 present typical irradiance—photosynthesis curves for species of the two main types. For the diatom it was unnecessary to present curves given per chlorophyll content, as this was identical for the algae grown either at a high or at a low illumination.

Whereas algal species of the *Chlorella* adaptation type, such as shown above, seem to be excellently fitted to living at weak irradiances, this does not especially seem to be the case for the species belonging to the *Cyclotella* type.

It has been shown that it is of decisive importance for algae living at strong irradiances to have a relatively high ratio between, on the one hand, the rates of enzymatic processes which correspond with a high content of photosynthetic enzymes, and, on the other hand, the rates of the photochemical processes, corresponding with a low content of photosynthetic pigments. In that way the effect of photooxidation is minimized. Physiologically it is, of course, of no importance if in "shade" algae the content of photosynthetic enzymes should be too high. However, in order to be able to compete with the other algal species found in habitats with low illumination rates, it must be absolutely necessary that the content of enzymes is not higher than necessary. At least 50% of the organic matter in an ordinary, healthy planktonic alga is protein, the major part of which constitutes enzymes. In the next chapter this important phenomenon will be discussed in more detail.

Algae adapted like *Chlorella* and found in the lowest part of the photic zone must be expected to have generally a higher rate of photosynthesis per cell at weak irradiances than surface algae. This is due to the generally higher content of pigments. At the same time, the rate of respiration per cell is lower. As a combination, therefore, the compensation point is found at a weaker irradiance. Whereas the compensation point in *Chlorella pyrenoidosa*

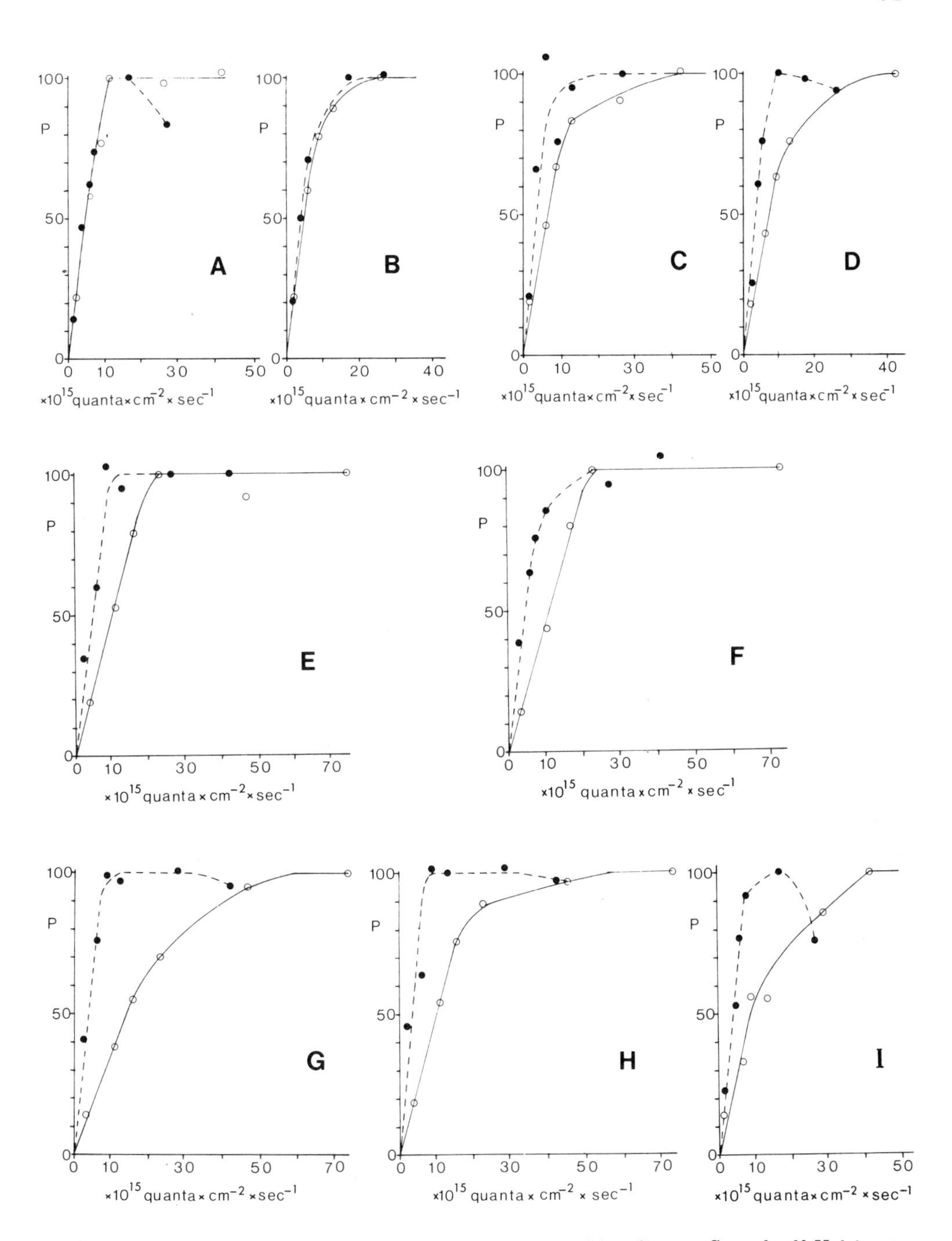

Fig. 37. Net photosynthesis (relative) as a function of irradiance, Sound off Helsingør, Denmark. ○: surface water; ●: water from the depth where 5% of green surface light was measured. A: December; B: February; C: March; D: April; E: May; F: June; G: July; H: October; I: November (after Steemann Nielsen and Hansen, 1961).

is found at about $0.5 \cdot 10^{15}$ quanta when grown at a very weak irradiance, it is found at about $2 \cdot 10^{15}$ quanta when grown at the strong irradiance of 36 $\cdot 10^{15}$ quanta. Due to the shade adaptation, algae in nature are able to grow at irradiances of only one third to one fourth of those which would be necessary if the algae were more or less sun-adapted.

In Fig. 37 curves according to Steemann Nielsen and Hansen (1961) are presented showing the rate of photosynthesis as a function of irradiance for plankton collected in The Sound off Helsingør (Denmark). The plankton was collected once every month over a period of a year both from the surface and from the lower part of the photic zone (5% of the green light at the surface). The temperature during the experiment was the same as found in the sea at the definite depths.

Fig. 38 presents the I_k of all the surface curves from all months as a function of the number of days before or after the summer solstice. At the same time before and after, the conditions for irradiance may be considered to be essentially the same. The values of I_k fall on two definitely separate curves, one for those before and one for those after the summer solstice. This indicates that the light adaptation of the algae is not due exclusively, although preferably, to the light conditions. The temperature has also some influence. When the length of the day is the same, in spring and autumn, the temperature at the surface is much lower at the former season. For example, on two days for experiments — February 3 and October 31 — that is, either 125 days before or 120 days after the summer solstice, the surface temperatures were 1°C and 12°C, respectively.

The curves for the plankton from the lower level (see Fig. 37), were all alike and — due to the halocline (cf. p. 114) — typical of "shade" plankton. The shape of the surface curves varies from typical "shade" curves in winter to typical "sun" curves in summer.

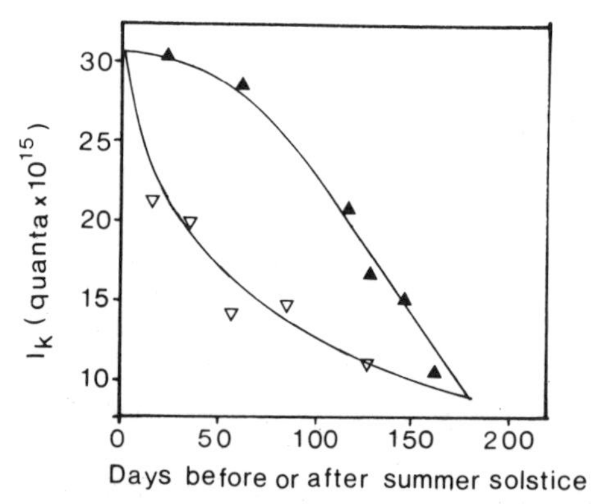

Fig. 38. I_k of the surface plankton from The Sound off Helsingør as a function of days from summer solstice. ▽: before, ▲: after summer solstice (after Steemann Nielsen and Hansen, 1961).

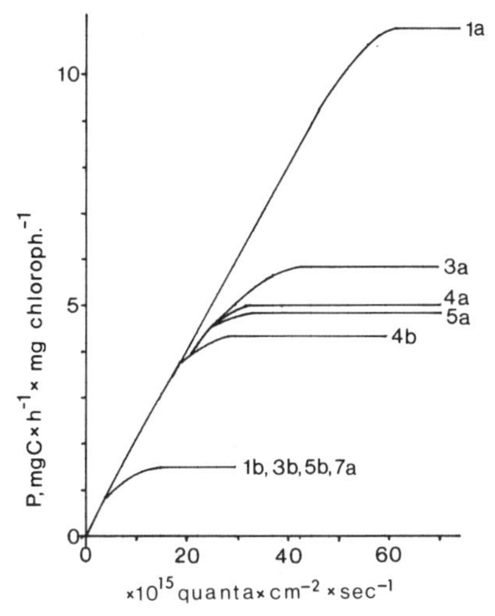

Fig. 39. Rate of gross photosynthesis (per unit of chlorophyll *a*) as a function of irradiance. *a*: surface plankton; *b*: plankton from the depth, where 1% of the green surface light was found. *1*: tropical plankton; *3*: temperate summer plankton; *4*: plankton from a northern area with very slight vertical stabilization; *5*: Arctic summer plankton; *7*: temperate winter plankton (latitude 56°N). Temperature: that of the habitat. (After Steemann Nielsen and Hansen, 1959.)

In Fig. 39 a series of schematic irradiance—photosynthesis curves is presented covering all kinds of marine habitats (see also Fig. 51A, B).

If we take an ocean where 1% of the surface light is found at a depth of 100 m, the depth of the photic zone will be just about 100 m. However, if the algae were not shade-adapted in the lower depth of the photic zone, the photic zone would be found only down to a depth where 3—4% of the surface light reaches. The photic zone would then be only about 75 m deep instead of the normal 100 m.

If the number of algae is more or less the same throughout the whole photic zone, the importance of their shade adaptation at the bottom of the photic zone is relatively slight for the production of organic matter calculated per surface unit of the sea or lake. Only a minor part of the total production is due to the shade-adapted algae in the lower part of the photic zone; cf. Ichimura et al. (1962) and Talling (1966). The situation is quite different, however, if the bulk of the algae is found in the lower part of the photic zone, a situation quite frequently encountered in the sea, e.g. in the Danish waters, as shown by Steemann Nielsen (1964a).

Summarizing, we may conclude that shade adaptation of the kind found in *Chlorella* is always of decisive importance for the algal population found in the lower part of the photic zone — and even for the population found near the surface in winter at higher latitudes. However, only in special cases is it of really decisive importance for the integral primary production per

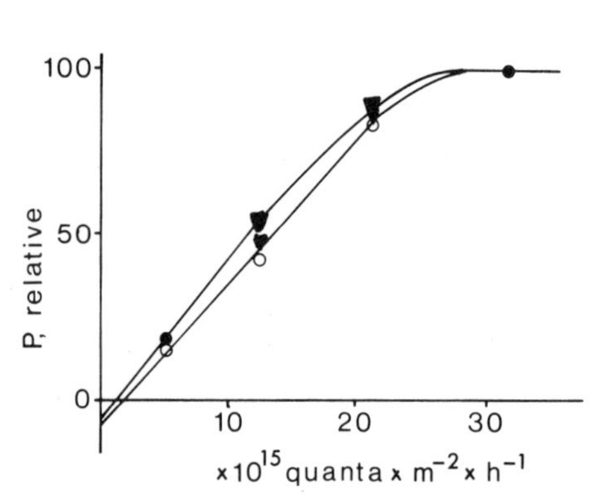

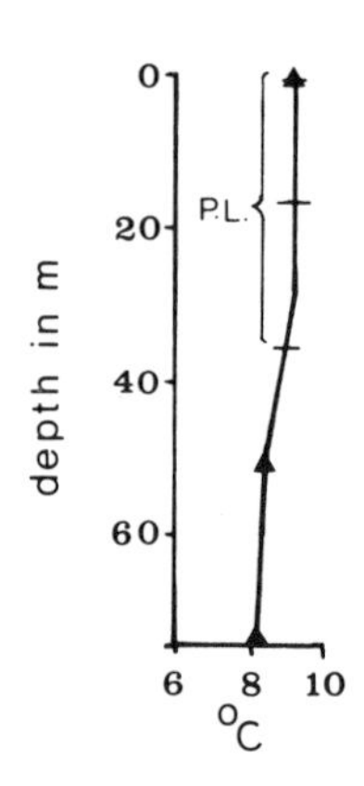

Fig. 40. The rate of net photosynthesis as a function of irradiance. ○: surface plankton; ▼: plankton from the depth where 10% of the green surface light was found; ●: plankton from the depth where 1% of the green surface light was found. To the right: the temperature as a function of the depth. Near the Faroes in July. (After Steemann Nielsen and Hansen, 1959.)

unit area. The sun adaptation of the surface plankton must always be considered of decisive importance for the integral primary production. When vertical stability of the water masses of the photic zone is lacking, the algae from all depths are nearly fully sun-adapted, as was shown by Steemann Nielsen and Hansen (1959) for a station over the submarine ridge between the Faroes and Iceland in July (Fig. 40; see also Fig. 51A).

One advantage of the adaptation of the algae to a strong irradiance — and probably the most important one — is their ability, under such conditions, to resist very strong irradiances better. If the dark reactions in photosynthesis are unable to keep pace with the photochemical reactions, the latter may induce photooxidations which again may destroy compounds in the chloroplasts such as the enzymes and thus decrease the overall rate of photosynthesis or even kill the algae. The lower the I_k , the weaker is the irradiance at which photooxidation can be expected.

In several bacteria it is evident that carotenoids function as photochemical buffering agents (cf. Burnett, 1965). A similar system is proposed by Krinsky (1964, 1966) for higher plants and green algae. The carotenoid pair, antheraxanthin—zeaxanthin could fulfill the requirements of being a "chemical buffer" to protect the cells against lethal photosensitized oxidations.

TEMPORARY CHANGES WHEN ADAPTING TO A NEW IRRADIANCE

In the first part of the chapter we have discussed the behaviour of algae which had grown for a long time under stable conditions at a definite irradiance. The adaptation to irradiance will change if we transfer the algae from one irradiance to another.

When changing from a weak irradiance to a really strong one, some temporary changes may take place in some species of planktonic algae. In *Chlorella*, for example, a substantial part of the photochemical mechanisms is inactivated and the rate of light-saturated photosynthesis decreases, probably due to photooxidative decomposition of some of the enzymes active in photosynthesis. Steemann Nielsen (1962) has treated these problems in a series of experiments with *Chlorella*.

Fig. 41 illustrates such an experiment. *Chlorella vulgaris* was grown in continuous incandescent light at $7.5 \cdot 10^{15}$ quanta. Curves showing the rate of photosynthesis as a function of irradiance were made: (a) before a transfer to $75 \cdot 10^{15}$ quanta; (b) after 3 hours at this irradiance; and (c) after a subsequent hour in the dark. The duration of each individual experiment was 15 min. After 3 hours at the strong irradiance, both the rate of the photochemical partial processes — determining the initial slope of the photosynthesis curve — and the rate of the enzymatic processes — determining the rate at light saturation — were depressed. In the dark a reactivation of both processes took place.

The inactivation of the photochemical mechanism does not involve any real destruction of chlorophyll, but seems to effect a protection against "surplus light energy", which otherwise could be used for photooxidation. The mechanism producing inactivation of the photochemical reactions in photosynthesis has the effect that, in the sea, curves showing the rate of photosynthesis versus depth on bright days usually have the maximum not at the very surface, but at a depth at which about 30—50% of the surface illumination is found. This is the case in the tropics and the subtropics and during the summer season also at higher latitudes (cf. Fig. 42, curve *a*). On dull days on the other hand, the highest rate of photosynthesis is found at

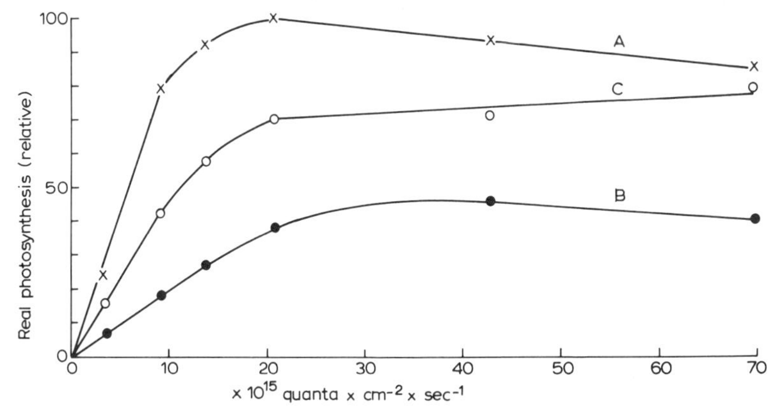

Fig. 41. The rate of photosynthesis (relative) as a function of irradiance for *Chlorella vulgaris* grown at $7.5 \cdot 10^{15}$ quanta, 20°C. *A*: directly from $7.5 \cdot 10^{15}$ quanta; *B*: after 3 hours at $75.0 \cdot 10^{15}$ quanta; *C*: after a subsequent hour in the dark (after Steemann Nielsen, 1962).

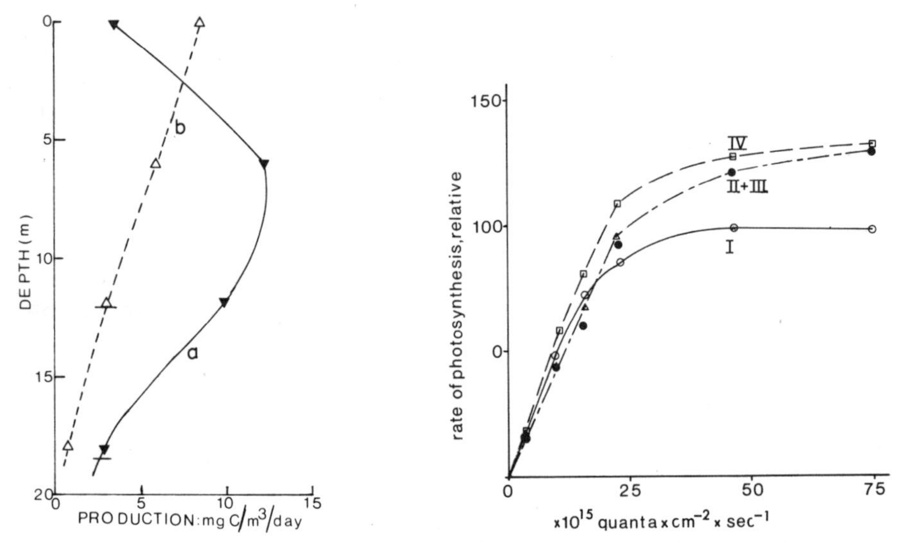

Fig. 42. The rate of gross production per m^3 per day as a function of depth, Great Belt at the end of April. *a*: *in situ* experiments (bright day); *b*: calculated for a very dull day (after Steemann Nielsen, 1964b).

Fig. 43. The rate of photosynthesis (relative) as a function of irradiance for *Cyclotella* grown at $7.5 \cdot 10^{15}$ quanta, 21°C. *I*: after 3 hours at $75.0 \cdot 10^{15}$ quanta; *II + III*: after 3 hours at $150 \cdot 10^{15}$ quanta; *IV*: after 3 hours at $250 \cdot 10^{15}$ quanta (after Jørgensen, 1964).

the very surface, as illustrated in Fig. 42, curve *b* (for more details, see p. 104).

Jørgensen (1964) has shown that the inhibition of the light-saturated rate of photosynthesis is not so pronounced, or is not found at all, in the species belonging to the *Cyclotella* adaptation type (see Fig. 36). When *Cyclotella* grown at $7.5 \cdot 10^{15}$ quanta was transferred to an irradiance 10 or 20 times as high, instead of an inactivation of the photochemical mechanism, by contrast the light-saturated rate of photosynthesis increased (Fig. 43). Thus, these species are better able to withstand a shift to a higher irradiance than cells of the *Chlorella* type.

THE TIME COURSE IN THE ADAPTATION TO A NEW IRRADIANCE

We will now discuss the development of the permanent physiological state of the algae due to transfer to a new irradiance. When shifting from a strong to a weak irradiance, no temporary changes seem to take place.

Curves presenting the rate of photosynthesis as a function of irradiance in *Chlorella vulgaris* were made by Steemann Nielsen et al. (1962) at different

TABLE VI

Average cell size and content of chlorophyll per 10^9 cells, *Chlorella vulgaris* (20° C)

Illumination	Average diameter (μ)	mg chlorophyll ($a + b$) per 10^9 cells
$75 \cdot 10^{15}$ quanta, steady conditions	5.6	0.40
$75 \cdot 10^{15}$ quanta, 24 hours after shift from 7.5 quanta	5.6	0.39
$75 \cdot 10^{15}$ quanta, 48 hours after shift from 7.5 quanta	5.7	0.40
$7.5 \cdot 10^{15}$ quanta, steady conditions	6.5	1.28
$7.5 \cdot 10^{15}$ quanta, 24 hours after shift from 75 quanta	5.4	1.09
$7.5 \cdot 10^{15}$ quanta, 48 hours after shift from 75 quanta	6.0	1.17

time intervals after the transfer from 7.5 to 75 or from 75 to $7.5 \cdot 10^{15}$ quanta. The rate of photosynthesis was measured per unit of cell numbers. In some of the experiments the concentration of chlorophyll ($a + b$) per unit of cell numbers was also measured.

Before presenting the photosynthesis curves after a shift from one irradiance to another, we shall discuss the variation in amount of chlorophyll per cell, the cell size and the actual growth rates during the first 24 hours after the shift in irradiance.

The data are presented in Tables VI and VII. As the temporary changes have already been discussed, no measurements for the first 17 hours after a shift from one irradiance to another will be described here.

Table VI shows that already 24 hours after a shift from a weak to a strong irradiance, the cells seem to be like those which had been growing constantly at the strong irradiance. The same fact is shown to some extent by the photosynthesis curve presented in Fig. 44. On the other hand, when shifting from the strong to the weak irradiance, a period of 24 hours or even 48 hours is not sufficiently long to produce the large cells characteristic of ordinary $7.5 \cdot 10^{15}$ quanta algae. The content of chlorophyll per unit of cell

TABLE VII

The factor (f) by which the cell number is increased during 24 hours, *Chlorella vulgaris* (20° C)

Illumination	f
$75 \cdot 10^{15}$ quanta, steady conditions	5.4
$75 \cdot 10^{15}$ quanta, first 24 hours after shift from 7.5 quanta	3.5
$7.5 \cdot 10^{15}$ quanta, steady conditions	2.1
$7.5 \cdot 10^{15}$ quanta, first 24 hours after shift from 75 quanta	2.2

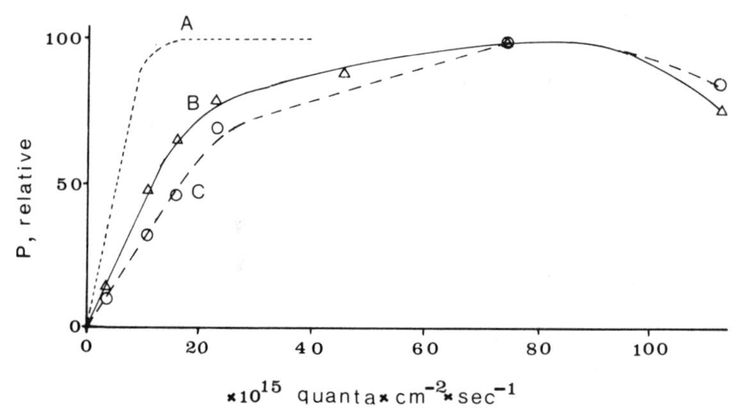

Fig. 44. The rate of photosynthesis (relative) as a function of irradiance. *A*: directly from $7.5 \cdot 10^{15}$ quanta; *B*: 17 hours; and *C*: 64 hours after transfer to $75.0 \cdot 10^{15}$ quanta. *Chlorella vulgaris*, 21°C (after Steemann Nielsen et al., 1962).

numbers and the photosynthesis curves (Fig. 45) after 24 hours at the weak irradiance are, however, nearly the same as those for cultures grown constantly at the weak irradiance.

The increase in cell numbers during the first 24 hours at the strong irradiance after a shift from the weak irradiance is intermediate between the increase at $7.5 \cdot 10^{15}$ quanta (steady condition) and $75 \cdot 10^{15}$ quanta (steady condition), showing that the growth rate during 24 hours had changed gradually from the originally low to the normally high rate found at $75 \cdot 10^{15}$ quanta. When shifting from a strong to a weak irradiance, the growth rate probably adjusts relatively rapidly to that characteristic for the weak one.

Fig. 44 shows the rate of photosynthesis as a function of irradiance, 17 hours and 41 hours after the transfer of algae from 7.5 to $75 \cdot 10^{15}$ quanta. Light saturation for both curves is put at 100. Light-saturated photosynthesis per unit of algae was about 10% lower 17 hours after the transfer as compared with 41 hours after. A slight difference is found between the two curves. The initial slope of the curve is somewhat steeper for the 17-hours curve. The I_k (cf. p. 53) for the 17-hour curve is $26 \cdot 10^{15}$ quanta, but for the 41-hours curve it is $33 \cdot 10^{15}$ quanta.

Fig. 45 presents, amongst other things, the rate of photosynthesis as a function of irradiance 17 hours and 41 hours, respectively, after the transfer from 75 to $7.5 \cdot 10^{15}$ quanta, the light-saturated rates in this case also being put at 100. The I_k was $14 \cdot 10^{15}$ quanta after 17 hours and 10 after 41 hours.

Fig. 46 shows the results of a series of experiments covering a period of 10 days. The irradiance was changed every third day from weak to strong or vice versa. The I_k is given as a function of time. The times for the changes in irradiance are indicated. The curve shows that a complete adaptation to the

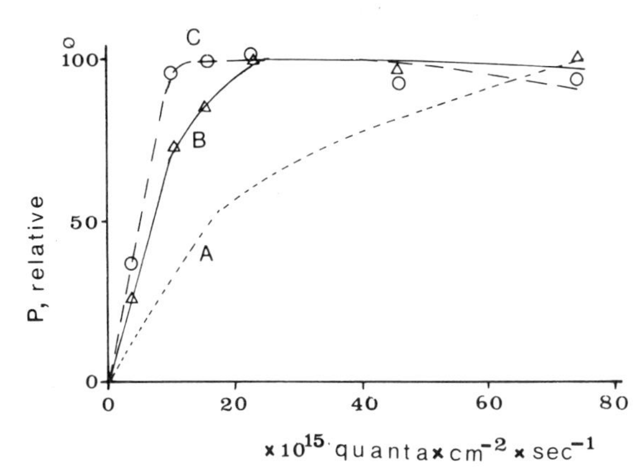

Fig. 45. The rate of photosynthesis (relative) as a function of irradiance. *A*: directly from $75.0 \cdot 10^{15}$ quanta; *B*: 17 hours; and *C*: 41 hours after transfer to $7.5 \cdot 10^{15}$ quanta. *Chlorella vulgaris*, 21°C (after Steemann Nielsen et al., 1962).

new irradiance takes about 40 hours, but that already after 24 hours the adaptation is considerably advanced.

The changes within the algal cell during the course of the adaptation to a new irradiance take place simultaneously with the growth of the algae.

Chlorella vulgaris grown under the present conditions normally produces about four daughter cells. If an adaptation only takes place when new cells are produced, the cell number must have increased by a factor of about 4 before the adaptation process is completed. This is in principal agreement with the results of the experiments.

Fig. 47 illustrates photosynthesis curves at various times after the transfer of surface plankton to a depth where 5% of the irradiance at the surface was found. We observe a gradual adaptation to "shade" conditions.

Steemann Nielsen and Hansen (1961) have shown that the adaptation to

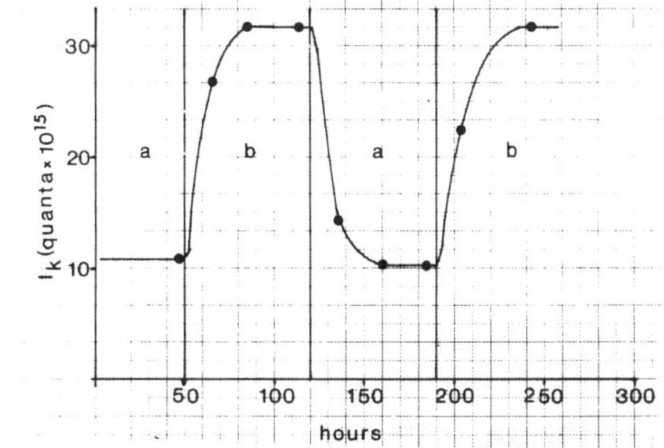

Fig. 46. I_k as a function of time after transferring *Chlorella vulgaris* from $7.5 \cdot 10^{15}$ quanta to $75.0 \cdot 10^{15}$ quanta, or vice versa; 21°C (after Steemann Nielsen et al., 1962).

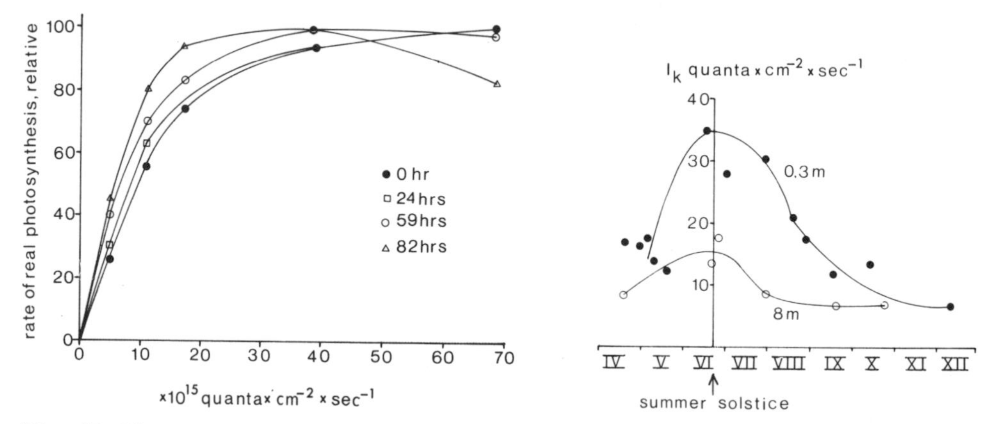

Fig. 47. The rate of photosynthesis (relative) as a function of irradiance. Surface water off Friday Harbour (August) and suspended for various times at the depth where 5% of the surface irradiance was found (after Steemann Nielsen and Park, 1964).

Fig. 48. I_k of microbenthic algae from the coast off Øresund (56° N) at various times before and after summer solstice (measurements from the topmost 2 mm of the sediment) at the temperatures at the sampling sites. ●: 0.3 m depth, ○: 8 m depth (after Gargas, 1971).

irradiance varies in a rather regular way throughout the year in phytoplankton collected near the surface in temperate waters (cf. Figs. 37 and 38). The experiments presented indicate that the different populations of algae found throughout the year are adapted to just the light conditions found during the last one or two days. No delay in adaptation takes place.

THE ADAPTATION IN BENTHIC ALGAE

The adaptation to irradiance in microbenthic algae has been investigated throughout a year by Gargas (1971). Two localities in the Øresund were chosen, one at a depth of only 0.3 m and one at a depth of 8 m. Fig. 48 presents the results. At a depth of 0.3 m, I_k varied between $36 \cdot 10^{15}$ quanta in June and $9 \cdot 10^{15}$ quanta in December. At a depth of 8 m the highest value of I_k was found in June: $17 \cdot 10^{15}$ quanta. The lowest found was $9 \cdot 10^{15}$ quanta. The microbenthic algae thus behave very much like the planktonic algae. The photic zone is only about 3 mm deep in the sediment, but during storms algae may be buried rather deep in the sand. Such algae from a depth of 6—7 cm have an I_k similar to that of the algae at the very surface of the sediment. No shift in I_k takes place during the burial deep in the sand, although the algae may stay in the dark perhaps even for a month.

Montfort (1936) showed that in macrobenthic algae a really strong irradiance generally increased the rate of photosynthesis in the species near the surface. However, the opposite was generally found in the algae normally growing at weak irradiances. As definitely pointed out by Montfort, in these algae we may have both a genetic and a physiological adaptation. He has finally shown that several "shade" species suffer direct harm when they are transferred to the direct sunlight at the surface. The depression of photosynthesis is often irreversible.

In thick algal thalli we may state that a possible physiological adaptation to various irradiances must take place by means of varying the content of the photosynthetic enzymes. As mentioned on p. 57 concerning the leaves of terrestrial plants, chlorophyll is normally found in considerable excess both in "sun" and "shade" leaves. The same must be the case in thick algae.

CHROMATIC ADAPTATION

In the present chapter we have, for the time being, only considered the adaptation to the quantity of light. We must of course also take the quality into consideration. Chromatic adaptation was much discussed at the end of the last century. Chromatic adaptation versus adaptation to the quantity of light gave rise to much discussion. The main advocate for the first point of view was Engelmann and for the latter it was Oltmanns (cf. Gessner, 1955). As genetic and physiological adaptation were to some extent intermixed, and because both points of view have later been shown to be more or less correct, we do not need to go into detail.

Gaidukow (1902) cultivated two *Oscillatoria* species (blue-green algae) in light of different colours and produced thereby algae of different colours. He was further able to state that the absorption of the light by the algae was highest in the wavelength region predominant in the light used for the cultivation of the algae. Harder (1923) was able to show that the rate of photosynthesis of the blue-green alga *Phormidium foveolarum* in blue light was considerably higher than in red light, if the alga had grown in blue light. If the alga had grown in red light, the opposite was found.

The problems concerning the physiological adaptation to the quality of the light are rather complicated as is presented by Egle (1960) in a review. As first shown by Haxo and Blinks (1950), absorption spectra and action spectra do not always give a true picture of the importance for photosynthesis of the various wavelengths of the incident light (cf. Haxo and Blinks, 1950; Yocum and Blinks, 1954). Light absorbed by phycoerythrin in red algae was found to be as effectively utilized in photosynthesis as that absorbed by chlorophyll in green algae, whereas light absorbed directly by the chlorophyll *a* of red algae appeared to be poorly utilized for photosynthesis at all wavelengths (see also p. 39).

Yocum and Blinks (1958) have shown that some red algae are able to adapt their photosynthetic mechanism to utilize red light with greater effectiveness, if they have grown for a week in red or blue light. The chlorophyll seems to be activated in the red part of the spectrum but not in the blue part. The "red-adapted" plants become de-adapted if they receive bright green light for a few hours.

CHAPTER 11

PHYSIOLOGICAL ADAPTATION TO TEMPERATURE

PLANKTON ALGAE

The adaptation to different temperatures is closely related to the adaptation to irradiance. In some cases we could even postulate that the shade adaptation of the algae in the lower part of the photic zone is only apparent; the difference in the shade of the irradiance—photosynthesis curve, and thus I_k, being only a direct consequence of the different temperatures in the upper and the lower part of the photic zone. Saijo and Ichimura (1962) have presented a striking example from the Oyashio area near Japan. The temperature at the surface was 20°C, and that at a depth of 20 m was 8°C. Photosynthesis curves made at the temperatures found at the depths in question gave an I_k of 12 klux (about $20 \cdot 10^{15}$ quanta) at the surface and of 6 klux (about $10 \cdot 10^{15}$ quanta) at 20 m. However, if the surface plankton was also measured at 8°C, I_k was 7 klux (about $12 \cdot 10^{15}$ quanta). We have seen exactly the same for plankton from the Danish waters (Steemann Nielsen and Hansen, 1961). The curves for surface water and subsurface water on June 4th, 1959, off Helsingør (cf. Fig. 37), showed a typical difference: I_k at the surface was about $16 \cdot 10^{15}$ quanta and at the 5% light depth (14 m) about 7.5 quanta. The temperatures were 14.3°C and 6.0°C, respectively. If the experiments had been made at the same temperature, the curves would have been rather similar.

However, the difference in I_k and hence also in the state of adaptation, is also found when the temperature is the same throughout the whole photic zone or even higher in the lower part of the photic zone. We always have a pronounced halocline in the Danish waters which prevents a real mixing from taking place throughout the photic zone, even during the periods when the temperature of the surface water decreases. Fig. 37 I (measurements on November 19th, 1959) thus shows a striking difference in I_k for the plankton from the surface and at a depth of 18 m in the lower part of the photic zone. The values are about $18 \cdot 10^{15}$ quanta and about $5 \cdot 10^{15}$ quanta, respectively, although the temperature was considerably higher at a depth of 18 m (11.9°C) than at the surface (7.9°C).

In Fig. 49 normalized curves (cf. Chapter 9) show the rate of photosynthesis as a function of irradiance in Arctic plankton during summer. Although the temperature difference between the surface (6°C) and at a depth of 50 m (2.5°C) was relatively insignificant, we found a typically dark-adapted plank-

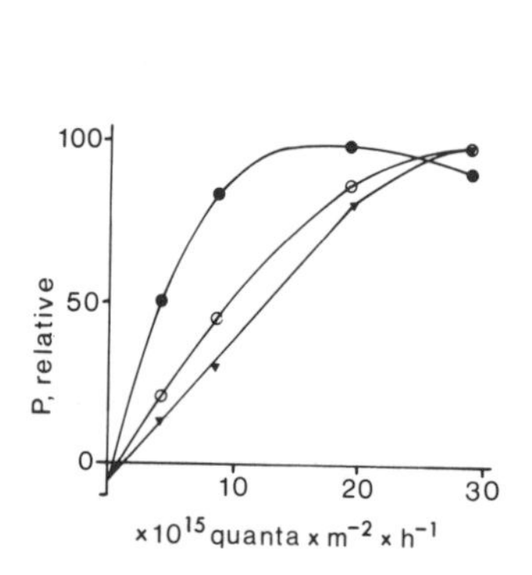

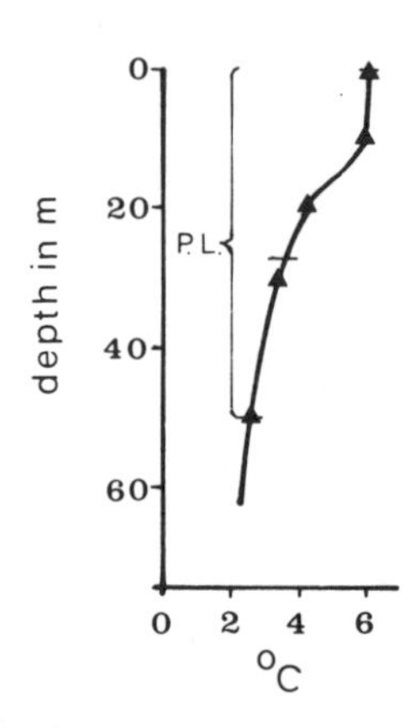

Fig. 49. Rate of apparent photosynthesis as a function of irradiance in Arctic summer plankton (64°N). ▼: surface; ○: a depth of 27 m; ●: a depth of 50 m. Temperature as that at the depth in question. To the right: temperatures as a function of the depth (after Steemann Nielsen and Hansen, 1959).

ton at 50 m, and a typically sun-adapted plankton at the surface. The I_k was found at 8 and 24 · 10^{15} quanta, respectively.

From experiments with natural plankton it appeared (Steemann Nielsen and Hansen, 1959) that Arctic surface plankton (64° N; Fig. 49) and temperate surface plankton (51° N; Fig. 50) which had been growing at approximately the same irradiance, but with a difference in temperature of 10° C, had relatively concordant values for I_k. For the Arctic plankton I_k was 23 · 10^{15} quanta, for the plankton from 51° N it was 28 · 10^{15} quanta (cf. further p. 73). In laboratory experiments with cultures of the diatom *Skeletonema* about the same dependence on temperature is found (see below).

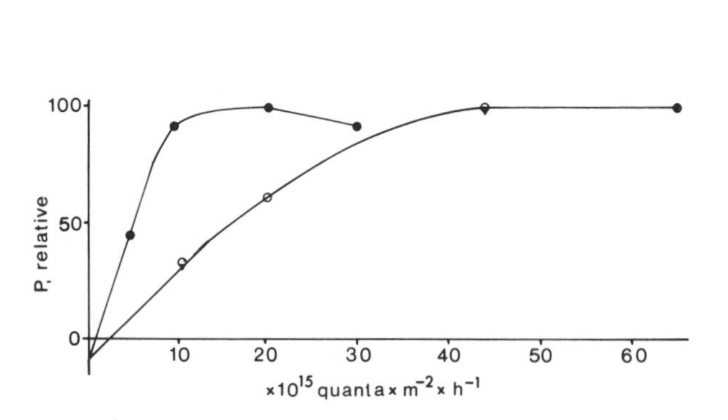

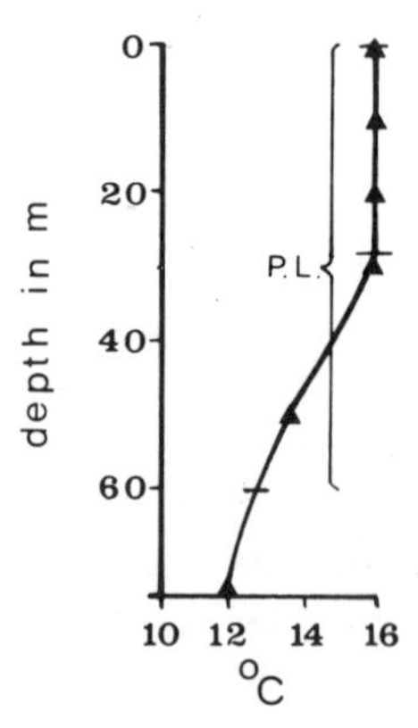

Fig. 50. Rate of apparent photosynthesis as a function of irradiance in temperate oceanic summer plankton (51°N). ▼: surface; ○: a depth of 28 m; ●: a depth of 60 m. Temperature as that at the depth in question. To the right: temperature as a function of the depth (after Steemann Nielsen and Hansen, 1959).

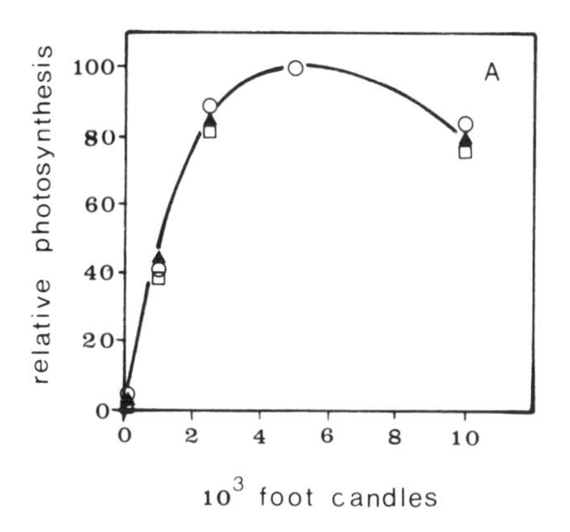

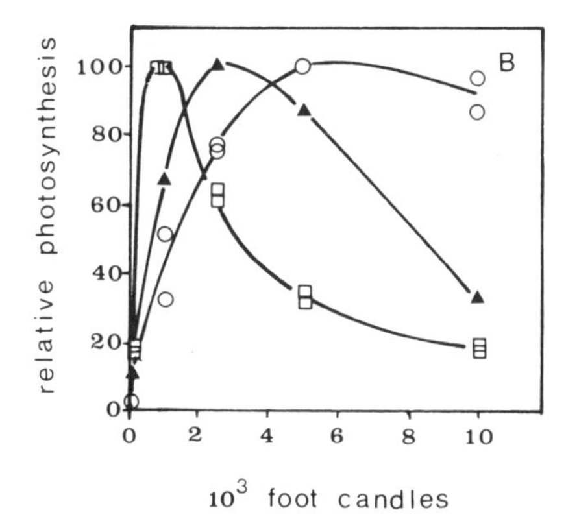

Fig. 51. Rate of photosynthesis as a function of irradiance in the subtropics off Bermuda. A. From November (no thermocline). B. From October (thermocline present). ○: surface; ▲: the depth to which 10% of the surface light penetrated; □: the depth to which 1% penetrated (after Ryther and Menzel, 1959).

Finally, in Fig. 51A and B, two series of irradiance—photosynthesis curves are presented from the subtropics — the sea off Bermuda, according to Ryther and Menzel (1959). In October, when the water masses were stratified (Fig. 51B), I_k at the depth to which 1% of the surface light penetrated was about 300 foot candles (= about $5 \cdot 10^{15}$ quanta) and thus very near to that found at the same light depth at higher latitudes. On the other hand, I_k for the surface plankton was $3 \cdot 10^3$ foot candles (= about $50 \cdot 10^{15}$ quanta). The latter is in accordance with the high irradiance found.

During winter, when the thermocline breaks down, all algae within the mixed layer must be assumed to be exposed to approximately the same average light conditions. They are circulated rapidly enough to prevent their being adapted to the irradiance conditions at any depth. According to Fig. 51A, I_k was found at $2.1 \cdot 10^3$ foot candles (= about $35 \cdot 10^{15}$ quanta). The temperature in the whole mixed layer was about 23° C, whereas in October it was 28° C at the surface and 19° C at the 1% light depth.

THE TIME COURSE IN THE ADAPTATION TO A NEW TEMPERATURE

When discussing the influence of temperature on the photosynthesis — and respiration — of planktonic algae, the time course is extremely important. Immediately after a shift of the temperature, we observe a simple effect on all chemical processes, including the enzymatic processes in respiration and photosynthesis. In the latter process it is only the rate at light saturation which will be influenced, enzymatic processes here limiting the overall processes. At weak irradiances, where temperature-insensitive photochemical

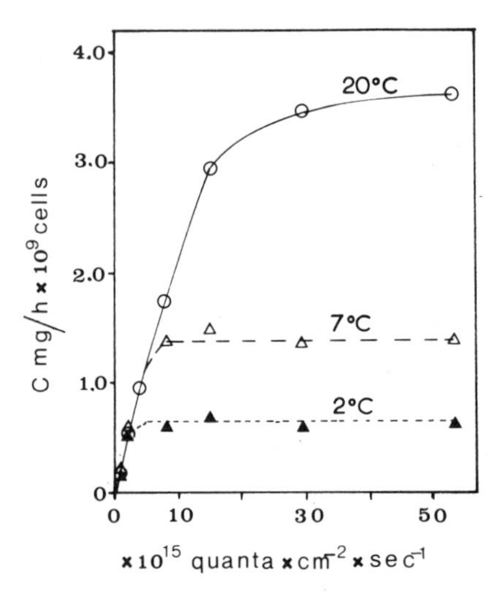

Fig. 52. The rate of photosynthesis as a function of irradiance in *Skeletonema costatum* grown at $7.5 \cdot 10^{15}$ quanta, 20°C, but transferred for 30 min to 20°C, 7°C and 2°C (after Steemann Nielsen and Jørgensen, 1968a).

part processes limit the rate of photosynthesis, temperature is of no importance. Several workers have convincingly shown this (e.g. Aruga, 1965). Fig. 52 presents some experiments with the diatom *Skeletonema costatum*.

However, for longer periods this simple situation will not last. The algae gradually adapt to the new situation. *Skeletonema costatum* was selected, because it is able to grow exceedingly well in nature in the whole temperature range 0—20° C. Therefore, it could be expected to be especially able to

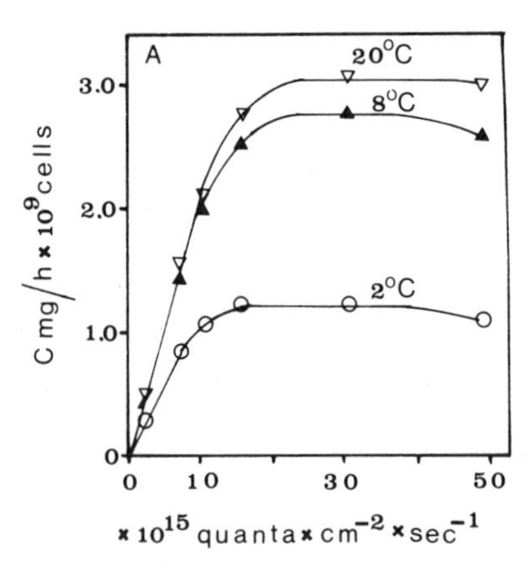

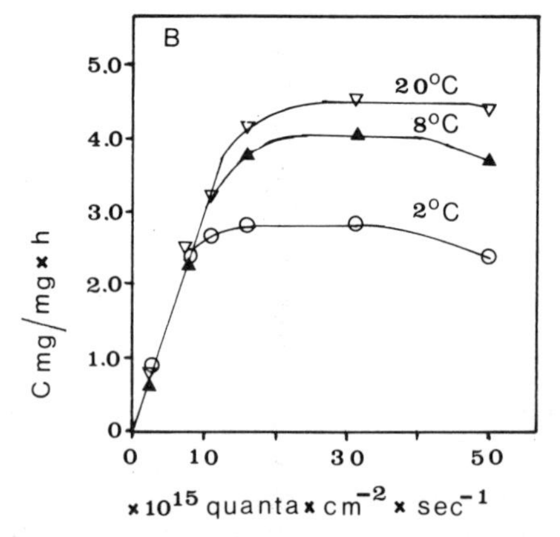

Fig. 53. The rate of photosynthesis as a function of irradiance in *Skeletonema costatum* grown in continuous light of $7.5 \cdot 10^{15}$ quanta at 20°C, 8°C and 2°C. The temperature during the experiment is the same as the growth temperature. A. the rate per cell number. B. the rate per mg chlorophyll (after Steemann Nielsen and Jørgensen, 1968a).

adapt to the various temperatures. The adaptation will ordinarily be accomplished in some few days. When working on the influence of temperature on the photosynthesis or growth of planktonic algae, for some species at least, it is practical to make the alterations of temperature in small steps, e.g. an alteration of about 5° C during each step (cf. Jørgensen and Steemann Nielsen, 1965). Otherwise the algae may suffer harm.

In Fig. 53A and B, the rate of photosynthesis of the diatom *Skeletonema costatum* grown in continuous illumination of $7.5 \cdot 10^{15}$ quanta at 20° C, 8° C and 2° C is presented as a function of irradiance. In Fig. 53A the number of cells is used as the unit, in Fig. 53B the weight of chlorophyll *a*. In the last case the initial slopes of all three curves are the same. However, if the number of cells is used as the unit, only the initial slopes of the curves for 20° C and 8° C are identical; a lower concentration of chlorophyll per cell at 2° C causes the less steep slope of this curve.

Whereas the light-saturated rate of algae, adapted to 20° C immediately after the transfer from 20° C to 8° C, decreases to about one third (cf. Fig. 52), after the termination of the adaptation back to 8° C the light-saturated rate is only slightly less than that at 20° C. The decrease by 12° C has, as shown above, the effect that the rate of enzymatic processes at constant enzyme concentration decreases to about one third. The rate of enzymatic processes is also dependent on the concentration of the enzymes, however.

ADAPTATION AND CONTENT OF PROTEIN

Steemann Nielsen and Hansen (1959) and Jørgensen and Steemann Nielsen (1965) have advanced the hypothesis that the concentration of enzymes per cell increases when the temperature is low. We must expect that *all* enzymes increase their concentration, thus not only those participating in photosynthesis. As repeatedly stressed, the rates of all processes taking place in an alga must correspond with each other. This was the easiest way of explaining the ability of the algae to maintain rates of enzymatic processes when these should have decreased due to the decrease in temperature. Further support for the hypothesis was the fact that the growth rate decreased pronouncedly when the temperature decreased, although the rates of photosynthesis and respiration did not decline. The enzymes must represent a major part of the protein in the cell. The protein constitutes, moreover, a considerable part of the total organic matter in a planktonic alga (at least 50%). Therefore, if the concentration of the enzymes increases — e.g. by a factor of about three — on lowering the temperature from 20° C to 8° C, much more organic matter has to be produced at 8° C in order to increase the number of cells by a factor of two; the growth rate thus automatically becomes lower, even if the rate of photosynthesis and respiration do not decrease.

The simplest way of proving the correctness of the hypothesis is to investigate the content of protein per cell number in algae adapted either to a high or to a low temperature. This was done by Jørgensen (1968). The amount of protein per *Skeletonema* cell was twice as high at 7° C as at 20° C.

According to experiments made by Morris and Glover (1974), it seems likely that not all species of planktonic algae are well suited for temperature adaptation.

As shown by Jørgensen and Steemann Nielsen (1965), the shift between a light period and a dark period is of decisive importance when *Skeletonema* is grown at a very low temperature. The photosynthesis curves presented in Fig. 53 were all made with algae grown in continuous light. However, if the algae at 2° C are grown in periods of 9 hours light ($7.5 \cdot 10^{15}$ quanta) and 15 hours darkness, the initial slope of the curve is exactly the same as the slopes of the curves at the higher temperatures. At the same time, the light-saturated rate of photosynthesis also increases (cf. Fig. 53 and 54).

It was further observed by Jørgensen (1968) that in *Skeletonema* cultures grown at a relatively low temperature (7° C) and continuous light, a few per cent of the cells were abnormal, being 4—5 times higher than the average cell height. This was not found in cultures grown in a light—dark regimen. Some of the cells in the culture in continuous light thus seem to be unable to start

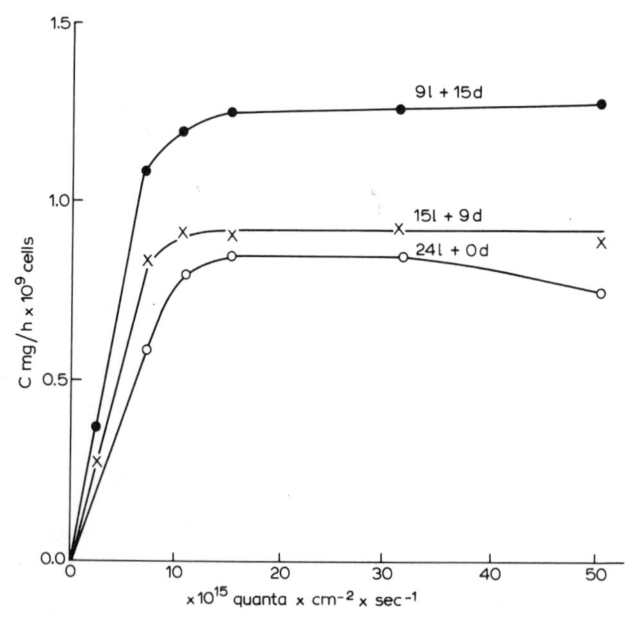

Fig. 54. The rate of photosynthesis in *Skeletonema costatum* as a function of irradiance. The algae were grown at $7.5 \cdot 10^{15}$ quanta given either continuously or in periods of 15 hours light and 9 hours dark, or 9 hours light and 15 hours dark, 2°C (after Jørgensen and Steemann Nielsen, 1965).

the cell division. This is further confirmation of the fact that the combination of a low temperature and continuous light is unfavorable for the growth of this particular diatom species.

In Danish waters *Skeletonema* is found as one of the most numerous plankton species throughout the year, thus also at low temperature in winter, when the days are short and the light intensities low. It has never been found in Arctic waters during summer, probably due to the fact that the days are too long (cf. Chapter 9 where the influence of the length of the day is discussed). We must, besides this factor, always also consider the irradiance and the temperature.

MACROBENTHIC ALGAE

Kniep (1914) investigated the influence of decreasing temperature on the rate of photosynthesis and respiration in marine macrobenthic algae. He found that the decrease in respiration was much more pronounced than that in photosynthesis. This was also confirmed by other workers. As they all seem to have employed rather weak irradiances, the result is easy to explain. The rate of photosynthesis has been more or less limited by the photochemical partial processes in photosynthesis which are independent of temperature (cf. p. 6).

Lampe (1935) investigated the rate of photosynthesis and respiration in the North European perennial brown alga *Fucus serratus* collected during winter. If the alga was transferred from the winter temperature of 5° C to 15° C and 21° C, he was able to show that both the rate of respiration and the rate of apparent photosynthesis at a strong irradiance immediately increased. This was in agreement with the fact that both processes were limited by enzymatic reactions.

The rate of apparent photosynthesis decreased at a weak irradiance with increasing temperature. This is due to the increasing effect of the temperature on the rate of respiration, whereas the rate of photosynthesis at a weak irradiance is not dependent on temperature. The temperature-independent photochemical reactions now limit the overall process. Lampe further investigated how the rates both of respiration and photosynthesis adapted, when he kept the algae for about a month at the higher temperatures. However, these investigations do not give a true ecological picture, because in nature an increase in irradiance follows the increase in temperature. The adaptation is both dependent on irradiance and temperature; but it is obvious that *Fucus serratus* is able to adapt.

As mentioned on p. 79, unicellular algae must divide — thus producing new cells — before becoming adapted to new growth conditions. In perennial macrobenthic algae, the cells in the thallus must be able to regulate their physiological adjustment.

Lampe also made experiments with the winter annual species *Porphyra hiemalis*. When transferred from 5° C to 21° C, it was unable to withstand the high temperature.

CHAPTER 12

THE MEASUREMENTS OF THE RATE OF PHOTOSYNTHESIS IN MARINE PLANTS

INTRODUCTION

The techniques used for measuring the rate of photosynthesis of the different kinds of life forms of marine plants vary greatly. It is relatively easy to measure the rate of photosynthesis in macrophytes if we measure it per unit of weight or per unit of surface area of a thallus or a leaf. However, if on the other hand, we wish to present the rate of photosynthesis per surface area of the sea we run into difficulties. The depth distribution of the plants is most unlikely to be homogeneous. At the same time, it is very difficult to get an impression of the illumination rates of the various parts of the plants. The results of such experiments can, therefore, in most cases only be rather rough approximations.

The estimate of the rate of photosynthesis of the microbenthos which is attached to rocks or stones also involves difficulties. Such work has, at least, been very much neglected, very likely due to technical difficulties. The microbenthos found on a soft bottom, especially of sand, has attracted the attention of scientists to a greater extent. We shall mention various techniques later in this chapter.

METHODS FOR MEASURING PHYTOPLANKTON PRODUCTION

First we shall turn to the planktonic algae which, as shown in Chapter 5, outside a narrow coastal fringe, are practically the only primary producers in the oceans. About 95% of marine primary production is very likely due to these minute algae.

The rate of photosynthesis in these plants may be measured per unit of their biomass or we may measure the rate per volume of sea water or below a certain surface unit. In cultures in the laboratory it is easy to determine the biomass of the algae. In nature, on the other hand, this is difficult. The particulate organic matter in the water is of very little use, because a major and varying part of it is dead. The total volume of the algae can also not be used. The vacuoles make up a varying and often major part of the total volume of these algae. Lohmann (1908) used a technique according to which he only measured the plasma volume of the algae. The technique is adequate, but unfortunately so time-consuming, that only very few workers have had the patience to use it. The only generally accepted technique of determining biomass is to measure the chlorophyll found in the algae. Unfortunately, the ratio between chlorophyll and biomass is rather variable — it varies at least by a factor of 5—10. The same seems to some extent to be the case concerning the relationship between ATP and biomass.

As measurements of the primary production by the planktonic algae have in most cases had the ultimate aim of presenting the background for estimating the total production in a sea area, it is easy to understand that they have mostly been done in a way giving the rate per unit surface area.

In most terrestrial vegetations and in marine vegetations of macrophytes, the production of organic matter is concentrated in units of plant material of a size large enough to make measurements relatively easy. In marine phytoplankton this is generally not the case. In the open ocean the minute planktonic algae are found distributed within a water layer of about 50—100 m deep, corresponding with the depth of the photic layer, the lower boundary of which is the depth where for 24 hours the rate of photosynthesis is the same as the rate of respiration. The amount of algae found in a single litre of water from the photic zone is therefore equal to the total amount of algae in this zone below a surface area of only about 0.1—0.2 cm^2. As the rate of production per day in the oceans is normally about 150 mg C/m^2, the average production per litre in the photic zone is only about 2 μg C per day, corresponding to about 6 μg O_2. This means that on average only 1/10,000 of the total carbon dioxide found in the water is assimilated per day, corresponding to 1/1000 of the oxygen. It is impossible to measure such small differences with chemical methods.

If the same amount of planktonic algae is concentrated in a shallow photic zone, e.g. only 10 m deep, such as is found in some coastal waters, a production rate of 150 mg C/m^2 per day can just be detected by means of oxygen determinations. However, in order to obtain more than a rather inaccurate measurement, a considerably higher rate of production must take place. Measurements by means of carbon-dioxide determinations are only possible in extreme cases.

Experiments involving the enclosure of water in bottles cannot be recommended for more than 24 hours. In many areas even considerably shorter experimental times must be used. Otherwise, e.g., the growth of bacteria may influence the result.

The main lines of the only two experimental techniques used in marine areas will be described: (a) the oxygen technique introduced by Gaarder and Gran (1927), often called the light-and-dark bottle oxygen technique; and (b) the ^{14}C technique introduced by Steemann Nielsen (1952). Both techniques measure the rate of photosynthesis. The experiments can thus be carried out identically, with the exception of the way in which the rate of photosynthesis is measured, either by determining the oxygen production or by estimating the uptake of the radioactive tracer, ^{14}C. In present-day research, practically only the ^{14}C technique is used, when working in the sea.

The oxygen technique

Water samples are collected from the various depths and siphoned into clear bottles and black bottles. One or several of the bottles containing water from the different depths are used for measuring the concentration of oxygen before the start of the experiment. The others are again lowered to the depths from which the samples came. Winkler titrations are used for measuring the concentration of oxygen.

Bottles wrapped in black material are suspended simultaneously with the ordinary clear ones. In this way it is possible to determine both respiration and photosynthesis going on in the water samples. The oxygen content in the black bottle, less than in the "initial" bottle, represents the rate of respiration, whereas the oxygen content in the clear bottle, less that in the black bottle, represents the rate of photosynthesis. It must, however, be mentioned that respiration includes not only that of the autotrophic algae, but also that of all other organisms, thus including that by bacteria and zooplankton.

The lower limit of the oxygen technique depends on the sensitivity of the Winckler technique. In a single titration it is impossible to determine oxygen with this technique to better than +0.02 mg O_2/l (by taking the average of 6 single titrations, $\pm$0.01 mg O_2/l). In many coastal waters it is even difficult or impossible to determine O_2 with this accuracy.

The ^{14}C technique

In the ^{14}C method, the incorporation of the tracer in the organic matter of the planktonic algae is used as a measure of the production. A definite amount of $^{14}CO_2$ is added to the seawater, the productivity of which is to be measured. The content of CO_2 (total) in this water must be determined or estimated. If we assume that $^{14}CO_2$ is assimilated by the planktonic algae only through photosynthesis and that $^{14}CO_2$ is assimilated photosynthetically at the same rate as $^{12}CO_2$, then by determining the content of ^{14}C in the plankton after the experiment, we also determine the total amount of carbon assimilated. It is only necessary to multiply the amount of ^{14}C found by a factor corresponding to the ratio between CO_2 (total) and $^{14}CO_2$ in the water at the beginning of the experiment. The amount of ^{14}C assimilated is determined by measuring the β-radiation from the plankton, which is retained by a filter with a maximum pore width of 0.5 μ. As the description of techniques is outside the scope of this book, only some few details will be mentioned. The manual edited by Vollenweider (1969, second edition, 1974) can, to some extent, be recommended. Geiger-Müller or liquid scintillation counting is used.

Filtering may be done either by suction or by pressure. The difference in vacuum or pressure must be less than 35 cm Hg.

The filters with the plankton are directly placed in a scintillation fluor after HCl treatment or they are placed in special holders for drying. The HCl treatment is done later by placing the filters for 5—20 minutes in a closed container above fuming hydrochloric acid. This is necessary in order to remove carbonate to be found, e.g., in coccoliths or in the filter. It is absolutely reprehensible to wash the filters with dilute solutions of HCl, such as is still done by some workers. This can lead to large losses of organic matter.

If the amount of originally bound ^{14}C in the plankton after an experiment is to give an absolute measure of the intensity of photosynthesis (the gross production), the following conditions must be fulfilled: (a) no $^{14}CO_2$ must be incorporated in organic compounds except through photosynthesis; (b) the rate of assimilation of $^{14}CO_2$ must be the same as that of $^{12}CO_2$; (c) no $^{14}CO_2$ must be lost by the respiration which takes place simultaneously with photosynthesis; and (d) no organic matter must be lost by excretion.

None of these conditions are, however, entirely fulfilled. Although the importance of each of them is only relatively slight at high light intensities, it is necessary to introduce corrections.

Dark fixation in planktonic algae is mostly about 1—3% of the fixation at optimal light intensity. In polluted waters, where huge quantities of bacteria are found, this may be of extraordinary importance. The same is the case near the lower boundary of the photic zone, where the rate of photosynthesis is low.

$^{14}CO_2$ is photosynthesized about 5% lower than $^{12}CO_2$. In light, about 50—70% of the CO_2 produced — or expected to be produced — by the respiration is, in some way or other, identical with the CO_2 fixed during photosynthesis, as was experimentally shown by Steemann Nielsen (1955). This is only of importance in experiments of relatively short duration. In ^{14}C experiments of very long duration, practically all CO_2 respired will be of the same kind as that assimilated. We now have static conditions. The ^{14}C technique, therefore, measures net production in such experiments.

Field measurements

For determination of the production of matter under 1 m^2 of surface, a method resembling Gaarder and Gran's oxygen method can be used. Instead of measuring oxygen metabolism, the intensity of photosynthesis is determined by the ^{14}C method.

The *in situ* method is time-consuming and therefore mostly too expensive to be used on expeditions with big ships. A simulated *in situ* method can be used instead. The samples from the various depths are placed in an incubator on the deck of the ship. In order to simulate the illumination conditions at the depths, plates of filter glass are placed above the bottles filled with subsurface water. In coastal waters, plates of neutral glass can be used with a relatively good approximation. In the open "blue" ocean, blue glasses must be used, such as was first done by Jitts (1963). To measure the illumination intensity both at the various depths from whence the water samples were collected, and below the filters in the incubator, a quanta meter must be used (cf. p. 13). On cruises another technique (physiological) may also be used (cf. p. 105).

The errors and the accuracy in the ^{14}C technique

The determination both of the counts of ^{14}C in the ampoules and in the organic matter of planktonic algae, must be made in such a way that they can be directly compared *inter se*. Most workers today prefer a kind of scintillation counting. Originally only Geiger-Müller counting could be used. Many complications arise both in Geiger-Müller counting and in scintillation counting. Common to all techniques is the fact that an amount of fixed ^{14}C is lost during ordinary drying of the filters, due to $^{14}CO_2$—$^{12}CO_2$ exchange. Wet filters, therefore, must be preferred in scintillation counting. It is, however, possible more or less to avoid the loss of volatile organic matter by formaldehyde vapour (Steemann Nielsen et al., 1974).

In Geiger-Müller counting we have, moreover, a loss of counts in the cases where the algal cells break up at the very end of the filtration and a part of the cell content is sucked into the pores of the filter. This loss can be determined by counting the filters both from above and below (Steemann Nielsen et al., 1974). Unfortunately, neither of the losses mentioned above are constant for the various species of planktonic algae. The biological technique for measuring the counts of the ampoules (Steemann Nielsen, 1965) in most cases corrected rather well for the two kinds of losses. Schindler et al. (1972) have outlined a scintillation technique without using filtration. After acidification and bubbling of the experimental water, 2.5 ml is placed directly in the vial.

Large and unnecessary errors seem often to have been introduced by several workers when using the ^{14}C technique. The following factors should be mentioned: (a) use of ampoules containing Cu or other poisons; (b) use of plastic water bottles which poison the water samples; (c) an acid wash of the filters instead of using hydrochloric acid vapour; (d) preservation of the samples with formaldehyde before filtration.

Even if all precautions are taken, we should not consider measurements of primary production to be correct by more than about ±30%. Considering the enormous variation of the rate of primary production in nature, this can be accepted.

METHODS FOR MEASURING BENTHIC MICROALGAL PRODUCTION

We shall not go into details, but only shortly mention the techniques used by marine workers (cf. further Vollenweider, 1974). Two in principle rather different techniques among others have been used to investigate the organic production by marine benthic microalgae found on sandy or muddy bottoms in shallow water. The first one was introduced by Pomeroy (1959). Changes in the dissolved oxygen concentration over light and dark, undisturbed sediment cores were used to estimate *in situ* primary production by benthic microalgae. A cylinder of glass closed on the top is placed over the surface of the sediment and left there for some hours or for half a day. Similar techniques have been used by Pamatmat (1968). The organisms are undisturbed during the experiments.

In another technique, originally introduced by Grøntved (1960), ^{14}C is used and the upper 3 mm (originally 10 mm) of the surface sediment is removed from a core and incubated under various light conditions in the laboratory. The resultant values are presented as the potential rate of production. The actual rate of production is tentatively suggested as one half of the potential rate at optimal light intensity. The same technique was used by Gargas (1970). A comparison of estimates of benthic primary production on a sandy beach measured by *in situ* oxygen and by laboratory ^{14}C methods, has shown that both methods gave rather similar measures of the magnitude of production (Hunding and Hargrave, 1973).

METHODS FOR MEASURING THE PRODUCTION OF AQUATIC MACROPHYTES

The annual production of most aquatic macrophytes is generally assumed to be estimated more or less satisfactorily by measuring the maximal seasonal biomass. If, apart from respiration, there are no losses of plant material between two sampling times — unfortunately often a rather unlikely situation — the net production, by definition, is equal to the observed change in biomass (Westlake, 1965). The major part of the estimates of the production of macrophytes in the sea has been obtained by means of such biomass studies. Wetzel (1965) and Vollenweider (1974) have reviewed the experimental techniques used for measuring the production of macrophytes.

Odum (1957) determined the metabolism of stream communities (freshwater) from the analysis of diurnal curves of dissolved oxygen. He investigated with the same technique the production of populations of the marine *Thalassia* in an area near Florida with strong tidal currents. The same technique has been used with more or less success by some few other investigators in other marine areas.

Changes in dissolved oxygen in transparent and opaque enclosures have only been used rather little for submerged macrophytes under field conditions (cf. Wetzel, 1964). The lack of an effective stirring of the experimental water first of all invalidates the measurements. The most frequently used method of investigating the photosynthesis and respiration of macrophytes has been to incubate portions of the plants in flasks and to determine the changes in oxygen concentration. It is also possible to apply a ^{14}C technique. Effective stirring can only be effectuated in laboratory experiments, but it has often been omitted even there. Removing the macrophytes, placing them into flasks, and returning them to their original depth imposes abnormal conditions that may seriously alter the resulting experimental values (according to Wetzel in Vollenweider, 1974). The worst source of error in plants like *Thalassia* with intercellulars seems to be gas storage within the leaves (cf. Zieman, 1973).

CHAPTER 13

THE UTILIZATION OF SUBMARINE LIGHT FOR PHOTOSYNTHESIS

INTRODUCTION

If we estimate the annual net carbon production of all seas to be $15-18 \cdot 10^9$ tonnes (see p. 121), in average only about 0.2% of the energy of the light (300—700 nm) penetrating the surface of all seas is transformed by means of photosynthesis into chemical energy. The main reason for the low rate of utilization is the fact that ordinarily only a very small part of the light energy is absorbed by the photosynthetic pigments in the marine plants.

Extremely high yields of photosynthesis and growth have been found in mass cultures of unicellular algae under natural irradiance, e.g. by Tamiya (1957). More than 10 g $C \times m^{-2} \times day^{-1}$ may be produced. It is important that the photic zone be very shallow. The light absorption by the water molecules is thus of very little importance. The same is true for organic and inorganic particles. Talling et al. (1973) have obtained equally high production rates per surface unit in two Ethiopian soda lakes having a depth of the euphotic layer of 0.6 m and a concentration of 200—300 mg chlorophyll *a* per m^2 in the euphotic zone. In marine coastal areas with a thin photic layer, about the same amount of chlorophyll has been found (Lorenzen, 1972). Ryther et al. (1971) found daily rates of up to more than 10 g $C \times m^{-2}$ in the Peru Current.

THE INFLUENCE OF THE SUBMARINE LIGHT

In the open oceans, but also in many coastal regions, the photic zone is relatively deep. The water molecules thus will absorb a considerable part of the light penetrating the surface; but also other causes are found for the absorption of light. We can mention minerals carried out from land or produced due to disintegration of coastal material. We may further add dead organic particles and dissolved organic matter, especially "yellow matter" either brought out from land with freshwater streams or found as remnants from an organic production which has taken place previously in the water.

In a theoretical ocean of absolutely pure water, 1% of the blue light (475 nm) and 1% of the green light (525 nm) would, according to N.G. Jerlov (personal communication, 1963), be found at about 160 and 90 m, respectively. In the tropics, subtropics and also during summer at somewhat higher

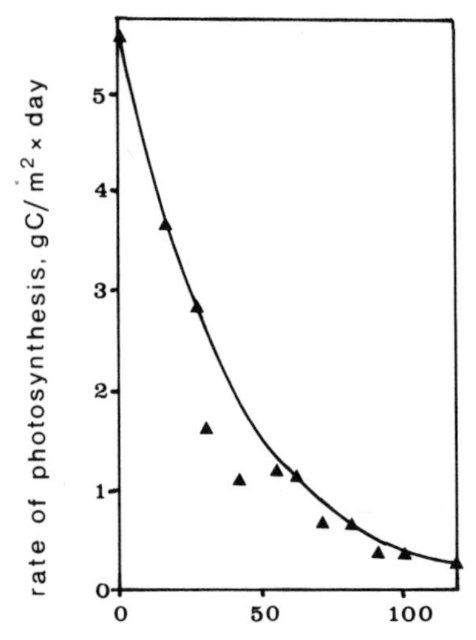

Fig. 55. Depth of photosynthetic layer and maximum rate of photosynthesis per m^2 surface (after Steemann Nielsen and Jensen, 1957, revised).

latitudes, the lower limit of the photic zone (cf. p. 20) is found at the depth, where about 1% of the blue + green light is found. In the theoretical ocean, this would be at a depth of about 140 m. This is not much deeper than in the most transparent parts of the real oceans. In the Sargasso Sea, for example, the depth of the photic zone is about 120 m.

Most of the light in these parts of the oceans is thus absorbed by the water molecules. Accordingly, only a very low standing stock of plankton can be found there. The rate of photosynthesis per surface unit must therefore be very low. The curve in Fig. 55 shows the maximum rate of photosynthesis versus the depth of the photic zone. All measurements from the Danish Galathea expedition were used for compiling the curve. The stations were grouped into 10-m classes according to the depth of the photic layer. From every class, the station showing the highest rate of production was selected. If the rate of production per m^2 is lower than the curve values, the opacity of the water must be considered to be primarily due to causes other than living planktonic algae.

Really transparent surface water seems only to be present in the centres of the eddies found in the subtropical parts of the oceans. In this "old"[1] and therefore oligotrophic surface water in the middle of the oceans, only few remnants from previous organic production are found. Much less transparent water is always found at higher latitudes, even during winter, when only extremely few planktonic algae are found. This is due partly to the age of

[1] By "old" surface water, we understand water which long ago ascended from deeper layers and now has been transported to other areas. The originally nutrient-rich "new" surface water has in the meantime become poor in these salts.

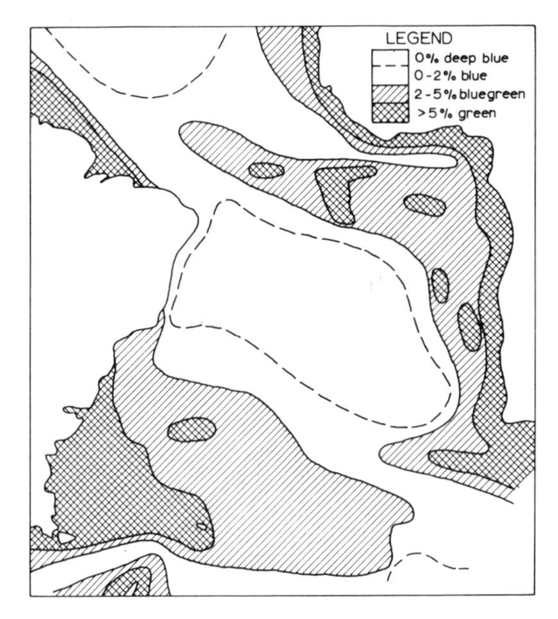

Fig. 56. Distribution of colour in the South Atlantic Ocean (after Schott, 1942).

the surface water as such. At higher latitudes, really "old" surface water is normally not found. The lower temperatures at the same time very likely also have some influence. The rate of the decomposition of organic matter is decreased.

Transparency is ordinarily not a very useful indicator of productivity in coastal areas. Here, the other factors besides the water molecules and the algae often play a major role in lowering the transparency. In the open ocean, on the other hand, the transparency — to be measured indirectly by means of the colour of the sea — presents a very useful indicator of the rate of production (see Fig. 56).

Platt (1969) has defined a coefficient k_b which measures the contribution of phytoplankton photosynthesis to the total optical attenuation coefficient. He suggests that k_b could be used as an index to compare the primary productivity of waters. As presented above, the use of k_b in coastal waters is often impossible.

When discussing the maximum rates of photosynthesis by the planktonic algae in the sea, we have to take into consideration that only at low irradiances are 8 quanta sufficient for the assimilation of a molecule of CO_2 (cf. p. 9). At a weak irradiance this means that about 25—40% of the light energy absorbed by the algae can be transferred into chemical energy (about 25% in blue light, about 40% in red light). As repeatedly mentioned, the overall rate of photosynthesis at strong irradiances is limited not by the photochemical reactions, but instead by the enzymatic processes. When the light saturation has first been reached with increasing irradiances, the utilization rate of the light energy therefore steadily decreases.

If light inhibition (cf. p. 58) sets in, the utilization is reduced further. In the photic zone in the sea, the utilization of the light energy absorbed by the algae varies considerably due, e.g., to the optical water type, the depth, the time of the day and the season. On average for the whole day and the whole photic zone — even if the planktonic algae are concentrated in a very shallow surface layer — only about a maximum of 10% of the absorbed light energy can be utilized in the photosynthetic process.

The different factors influencing the rate of primary production by the plankton in the sea, such as the replenishment of nutrients, light, grazing, etc., cannot be traced separately. They are generally involved in such a way, that all of them appear to act as limiting factors at the same time. Taking all seas as a whole, the replenishment of nutrients in the productive layers must normally be considered to be the essential factor that determines the magnitude of the annual organic production. Grazing is perhaps the most important factor determining the rate of production during any particular day. This is due to the fact that the size of the population of algae is governed to a high degree by the rate of grazing. During winter at higher latitudes, light is, of course, the very factor limiting the rate of primary production. But on each separate day during all seasons, light influences the amount of organic matter produced. We must definitely distinguish between momentary limitation of the organic production and the limitation for a year or a production season. The latter is adjusted due to the combination of all the factors responsible for the production of organic matter by the planktonic algae.

We must thus always consider underwater light when discussing organic production. On the other hand, we must emphasize that both the nutrient conditions and grazing also influence the light conditions below the surface by influencing the algal concentration in the water. This was clearly shown in Fig. 55.

In order to derive a widely applicable expression for the total or integral photosynthesis of a natural population beneath a unit area of surface, Talling (1957) used depth profiles obtained by suspending bottles containing the freshwater diatom *Asterionella formosa*. The integral photosynthesis was related, under a wide range of conditions, to the logarithm of the surface irradiance. In Talling's approach, the assumption of a homogeneous population is an important factor. This is mostly not the case in the sea. Furthermore, no consideration of inhibition in strong light near the surface was included. Steele (1962) has developed a theoretical photosynthesis—light relationship which includes the effects of inhibition in intense light. Photosynthesis is determined per unit chlorophyll and carbon, by C:chlorophyll ratio as function of incident radiation and nutrient concentration. The equations appear to describe the main trends in the open North Sea and in the Sargasso Sea. Becacos-Kontos and Svansson (1969) have found satisfactory agreement, when using Steele's theoretical relationship in the eastern Mediterranean.

THE INFLUENCE OF THE HYDROGRAPHIC CONDITIONS ON THE LIGHT CONDITIONS

The general hydrographic conditions in the sea also have a definite influence on the light conditions. A planktonic alga has to follow the water masses carrying it. The photic zone in the sea may either have a smaller, a similar or a greater vertical extent than the layer in which the water masses are more or less vertically mixed daily. The depth of the thermocline, or in some few cases the halocline, may vary considerably from one area to another without being identical with the depth of the photic layer. In winter at higher latitudes in most parts of the oceans, no thermocline at all is found. The wind will therefore often — although not every day — be able to mix the water down to depths of several hundred meters.

When the depth of the photic zone is more or less identical with the depth just above the thermocline, or in coastal areas above the bottom, the most simple situation is found for the light utilization of the planktonic algae. Both quantitatively, qualitatively and concerning the physiological adaptation, the algal population is identical. If the photic zone penetrates into the thermocline or even into the water masses below, the plankton population is heterogeneous both concerning quantity and state of adaptation.

In the area between the brackish Baltic and the saline North Sea, a strong vertical stabilization caused by the halocline is always found. The plankton is thus able to grow near the surface during the whole winter, if the sea does not freeze (Steemann Nielsen, 1964a) (cf. p. 114). If the stabilization is caused solely by a thermocline such as in the oceans, the latter will break down during winter at higher latitudes. The plankton algae are not able to grow, because the respiration rate in the whole mixed layer now exceeds that of photosynthesis. Some few specimens of algae are found in the water masses. They are either surviving by means of heterotrophy or they are found in a more or less dormant stage. When the thermocline is again established in spring, these algae are able to start growing again.

Recently several workers have considered the very important ecological problem concerning the dark survival of photoautotrophic microalgae (cf., e.g., Smayda and Mitchel-Innes, 1974). In summary, all observations and experiments suggest that the dark-survival potential varies between species, is dependent on temperature, and seems to be prolonged by periodic lighting at subcompensation intensities. The last phenomenon is very likely of decisive importance for the survival of the planktonic algae in the northern Atlantic at latitudes of about 60°N. Here the period of vertical mixing of the water masses down to several hundred meters lasts about six months!

Braarud and Klem (1931) were the first to realize the influence of a lack in stabilization of the water masses in the sea. A growing phytoplankton population can only occur if the vertical extent of the mixed layer is less than a certain maximum depth, called the critical depth. Sverdrup (1953) tackled the problem theoretically. Using certain assumptions, he was able to compute the critical depth and to follow the outburst of phytoplankton during spring in the Norwegian Sea. In its final state this took place during the second week of May.

Patten (1968) has questioned Sverdrup's theory. He has claimed that the algae theoretically should be able to adapt in such a way that net community production would be positive, even in a completely mixed water column. Although Patten's idea is tempting, it is without physiological support and hardly correct (cf. p. 52). At the temperatures found in the northern part of the North Atlantic during spring, planktonic algae are hardly able to adjust the rate of photosynthesis and respiration sufficiently to give rise to any growth in the vertically mixed water layer. This is about ten times deeper than the photic zone. The survival of the algae, such as mentioned above, is another matter. Furthermore, the start of the phytoplankton outburst in the northern part of the North Atlantic and in the Norwegian Sea closely follows a scheme which is in agreement with Sverdrup's theory.

Near the coasts, the shallow depths counteract the lack of vertical stabilization. The same is found over isolated banks in the oceans. The depth at the Faroe Bank in the North Atlantic, west of the Faroe Isles is only about 100 m. In the beginning of May 1934, a concentration of plankton algae of about 200 ml^{-1} was found here. In the ocean surrounding the Bank, the concentration of algae was still low — about 20 ml^{-1} (cf. Steemann Nielsen, 1935).

THE INFLUENCE OF DAY-TO-DAY VARIATIONS OF THE IRRADIANCE

Steemann Nielsen (1964b) investigated the influence of day-to-day variations, in the middle of April, in the irradiance of the surface of the sea in a temperate locality — the Great Belt, 55°N. It was shown that during this season, daily variations in the light penetrating the surface normally have only a relatively slight influence on the position of the compensation depth and on the rate of production at the very surface. During winter, on the other hand, day-to-day variations in light are of decisive importance. In this article, not only was the influence of the day-to-day variations in light in April on the variation in the total rate of production below a surface unit of the sea discussed, but also the author considered how the vertical distribution of the planktonic algae is of importance for the extent to which day-to-day variations in light influence the rate of production. It was shown that the highest rate of production is not always found on the brightest days (cf. Table VIII).

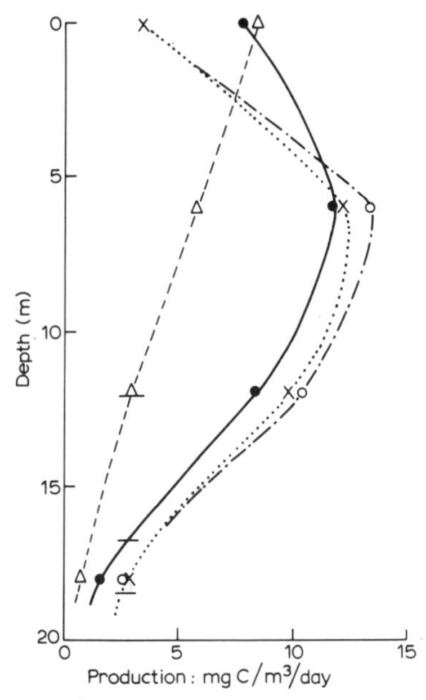

Fig. 57. The rate of gross production per m^3 per day as a function of the depth. (a) dotted line (×) according to the *in situ* experiments; (b) solid line (●) calculated for an average day; (c) dots and dashes (○) calculated for a bright day; and (d) dashed line (△) calculated for dark days. The position of the compensation depth is shown. Great Belt, 20th April (after Steemann Nielsen, 1964b).

An *in situ* station was maintained from astronomical noon to sunset. Measurements were made at 0, 6, 12, 18 and 24 m depths. At the same time, water samples from all of these depths were given $20 \cdot 10^{15}$ quanta incandescent light in a tank. The tank was cooled by running surface water. The vertical distribution of temperature is sufficiently homogeneous during this time of the year to permit such a procedure. Only small corrections for temperature differences had to be made.

In Fig. 57 the dotted curve (*a*) represents the *in situ* measurements, calculated as gross production for a whole day (20th April). The other curves are derived from the following measurements: (1) the above-mentioned tank measurements at $20 \cdot 10^{15}$ quanta; (2) curves showing the rate of photosynthesis as a function of irradiance in the plankton collected at different depths in April; (3) vertical penetration of light, measured by means of a submarine photometer; (4) irradiance at the surface for every hour of the day in April. The calculations are made for: (*b*) an average day in April (continuous line), (*c*) for the brightest days in April (dotted and dashed line) and (*d*) for the darkest days in April (dashed line) (cf. Fig. 4). Both on a very bright day and on an average day, light inhibition takes place at the very surface. For the very bright day, the decrease at the surface due to light inhibition was put at 56%. This is in accordance with the result of the *in situ* measurements on 20th April, which was sunny and with a partial coverage of white clouds only during a short part of the day. For an average day, the inhibition was tentatively put at 5%, in accordance with the results from similar experiments made in September on a day with average light conditions.

The curves presented in Fig. 57 show clearly that in April the rate of gross production measured per unit of surface area does not vary very much from a very bright day to a day with average illumination. On the very dark days, a reduction to about half the rate of production is found.

The actual *in situ* curve is very near to the curve for the brightest day, as also might be expected from the actual irradiances measured throughout the day of the investigation.

The tank experiments showed that the light-saturated rate of photosynthesis was fairly constant at most depths. Only at the very surface was the rate considerably lower. However, in the very strongly stratified Danish waters, the vertical distribution of phytoplankton often varies considerably. The same is also found in many areas in the oceans. Sometimes the main bulk of algae is found near the surface. In other cases, on the contrary, it is found near the lower boundary of the photic zone.

A different vertical distribution of the phytoplankton must theoretically be of considerable importance for the influence of day-to-day variations in the light conditions on the rate of production below the surface unit. It must be remembered that the rate of photosynthesis in the lower part of the photic zone is proportional to irradiance. At the surface, with the exception of the middle of winter, this is by no means the case. Due to light inhibition, the rate is even higher here on an average day than on a very bright day.

Hence, calculations were made, taking into account varying quantities in the vertical distribution of algae, using as a unit the light-saturated rate of photosynthesis. Otherwise, the results of the experiments made in the Great Belt on 20th April were used. The relative rates of gross production within the photic zone per m^2 of the surface are presented in Table VIII. The rate for an average day is always put at 100. In addition to the rates for the actual vertical distribution on 20th April (a), the rates are presented for: (b) an even vertical distribution of the light-saturated rate of the photosynthesis at all depths; (c) a distribution according to which this rate gradually increases from the surface to a depth of 18 m, overall by a factor of 8; and (d) a distribution where the rate instead gradually decreased by a factor of 8, from the surface to 18 m.

Table VIII shows that the rate of gross production per m^2 surface in April is practically the same on very bright days and on average days, if the phytoplankton is vertically evenly distributed. On the very dark days, the rate is reduced by less than 50%. If the main bulk of plankton is found near the lower boundary of the photic zone, example (c), the rate is three times

TABLE VIII

Relative rates of gross production per m^2 surface in the Great Belt in April (see text for explanation of c and d)

Vertical distribution of potential rate	Bright days	Average days	Dark days
(a) actual distribution (20 April)	112	100	53
(b) even vertical distribution	102	100	53
(c) 0 m = 1, 6 m = 2, 12 m = 4, 18 m = 8	121	100	40
(d) 0 m = 8, 6 m = 4, 12 m = 2, 18 m = 1	87	100	70

as high on very bright days as on very dark days. Finally, if the main bulk of plankton is concentrated near the surface as in example (d), the difference in the rate of gross production between very bright, average and very dark days is rather slight, the highest rate being found on an average day.

The transition area between the Baltic and the North Sea is characterized by an extremely pronounced vertical stratification. In several other coastal areas, very little stratification is found. However, here we cannot expect similar conditions with respect to day-to-day variation in the irradiance. In the open oceans, on the other hand, where the depth of the photic zone is normally considerable, the vertical stratification is pronounced, with the exception of upwelling areas. In the open oceans, we thus may expect that, in principle, the influence of day-to-day variations during summer is like that shown here for the Great Belt.

LIGHT UTILIZATION IN THE HIGH ARCTIC

Finally in this chapter, an enigma concerning the light utilization in sublittoral perennial algae in the high Arctic must be discussed. Lund (1959) and Wilce (1967) have found resident populations of perennial algae at depths exceeding 100 m. Only very little light energy can penetrate down to this depth in the high Arctic, where the transparency never seems to be really high and where surface irradiance is found only during a part of the year.

In such a locality — Mould Bay, N.W.T. of Canada, 75° N — Wilce showed that at a depth of 35 m, where a considerable amount of perennial algae were found, less than 0.1% of the surface light (440—580 nm) was measured, even during June and July. It was only in the second part of August that the irradiance at this depth exceeded 1% of that at the surface. The rate of organic gross production (= real photosynthesis) per year in such an algal population must very likely be exceedingly low. If the rate of respiration at the negative temperatures found in the habitat is also exceedingly low, theoretically it should be possible for such algae to exist. However, the growth rates consequently must also be exceedingly low. One hundred years here perhaps correspond with a single year at habitats outside the high Arctic! Due to the negative temperature, algae perhaps can tolerate such a slow growth rate. Wilce and Lund, on the other hand, suggest facultative heterotrophy as a means by which these algae adapt to their environment. This possibility cannot be denied.

However, it must be noted that for high Alpine lakes where throughout the year a considerable diversity of nannoplanktonic algae is found, Pechlaner (1971) suggests that, even during winter and spring, autotrophic rather than heterotrophic production dominates. In an additional discussion of Pechlaner's article, Fogg stated that it is well attested that all sorts of algae can take up organic substances at a very low light intensity, although not in the dark. This probably makes it possible to use light more efficiently, when the irradiance is very weak. The enigma concerning primary production in aquatic habitats with extremely low temperatures cannot at present be regarded as finally solved.

CHAPTER 14

THE RATE OF PRIMARY PRODUCTION IN THE SEA

PHYTOPLANKTON

In the years before the last World War, some few measurements (the first by Gaarder and Gran, 1927) were made in coastal waters concerning the rate of primary production by the planktonic algae present below a unit of the surface. The oxygen technique was used for measuring the rate of photosynthesis and respiration. However, it was only after the introduction of the ^{14}C technique during the Danish Galathea expedition (1950—52), that primary production could be measured in all kinds of waters. During the last 25 years, a multitude of measurements have been made, both in oceanic and coastal regions. At present we have a relatively good estimate of global production.

It is worth mentioning that Sverdrup (1955) was able, by means of theoretical considerations, to present a schematic representation of the probable relative productivity of all oceans. He assumed that productivity depends on the rate at which plant nutrients in the surface layers are renewed and that the renewal takes place due to upwelling and turbulent diffusion. In Fig. 58, a chart is shown presenting the annual rates of primary production in all seas, according to all measurements made up until now. This chart is, in principle, the same as Sverdrup's relative chart.

Sournia (1969) stressed that it is impossible to present any generalizations concerning the production in tropical waters. The same can, in fact, be said concerning temperate and Arctic (Antarctic) waters. In all temperature regions, areas with both high and low rates of primary production are found.

As discussed in Chapter 12, we cannot reckon with a high accuracy of measurements of the rate of primary production. Nevertheless, this is not very important, if global productivity is to be presented. This is due to the enormous variation, from area to area, in rate of production both per volume and per surface unit. Per $m^3 \times day^{-1}$ in the sea, gross primary production varies at least with a factor of 10^4, even if we disregard low measurements due to winter conditions, etc. The highest rate is at least $10 g C \times m^{-3} \times day^{-1}$. The rate of gross production per surface unit varies with a factor of about 10^2. It is thus easy to understand that discrepancies, even those as high as 50%, do not play a dominating role and do not spoil the general outlines.

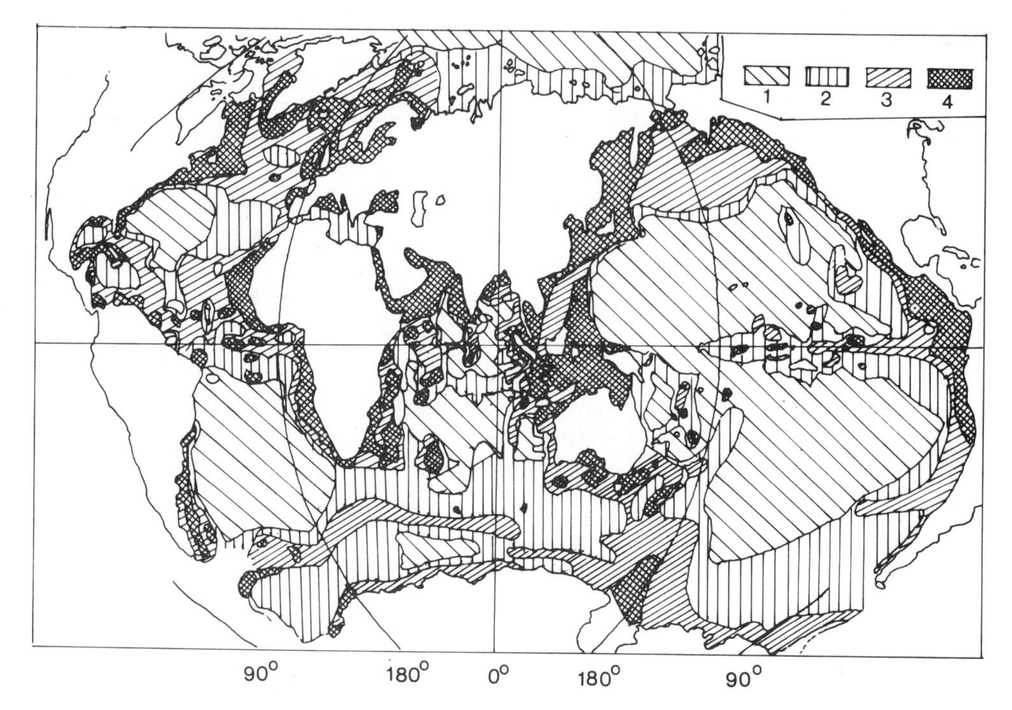

Fig. 58. Distribution of primary production in all oceans. Units are in mg C per m^2 per day. *1*: less than 100; *2*: 100—150; *3*: 150—250; *4*: more than 250 (simplified after Koblentz-Mishke et al., 1970).

The Atlantic Ocean

We shall start with the open Atlantic Ocean. Many measurements of primary production have been made there. It would serve no purpose to mention all investigations. Instead, only some of the most pertinent will be dealt with and the papers reviewing the whole ocean or parts of it will be presented. Many investigations have been made by Russian workers. Unfortunately for western readers, a very considerable part of their data has been published only in the Russian language.

The Atlantic Ocean will be discussed in more detail than the Indian and the Pacific Oceans. The main reason for doing so is that the Atlantic and the adjacent waters are by far the best-known sea areas with respect to the rate of primary production. However, the three oceans are rather alike concerning most, but not all, of the general trends.

The North Atlantic and the South Atlantic are both characterized by large circulating current systems. In the South Atlantic these run counterclockwise and in the North Atlantic, clockwise. In the South Atlantic the Benguela Current runs in a northerly direction along the west coast of South

Africa. It receives a material part of the water masses from subsurface layers. The upwelling of water is of particular importance close to the coast.

The South Equatorial Current crosses the South Atlantic in a westerly direction just south of the Equator. A part of the current runs into the North Atlantic to the west, combining there with water masses from the North Equatorial Current. The main part — the Brazil Current — continues in a southerly direction along the coast of South America. The central part of the South Atlantic at middle latitudes constitutes a large anticyclonic counterclockwise eddy.

A similar anticyclonic eddy, the Sargasso Sea, is found in the North Atlantic. In the Northern Hemisphere the direction is clockwise. The North Equatorial Current starts near Africa, where pronounced upwelling occurs off the coast. The current flows across the ocean and a part of it enters the Caribbean Sea. Another part is the Antilles Current running just east of the West Indies. North of the central eddy — the Sargasso Sea — the east-going Atlantic Current is found. Between the North Equatorial Current and the South Equatorial Current a countercurrent is found in the eastern part of the Atlantic just around the Equator. It runs in an easterly direction and gives rise to divergences bringing subsurface water to the surface. Below, some few comments will be presented concerning the hydrographical conditions in the adjacent seas — such as the Mediterranean.

As already theoretically predicted by Sverdrup (1953) and in close agreement with the earlier investigations of Lohmann (1920) and Hentschel (1933—1936) concerning the quantitative distribution of planktonic algae in the Atlantic, a close dependence of the rate of primary production on the hydrographical conditions is found. Although some areas are much better investigated than others, sufficient data are available, even on an annual basis, to describe the rate of primary production in the whole Atlantic.

To start with the anticyclonic eddy in the North Atlantic, the Sargasso Sea; the rate of primary production there was investigated at an early stage (Steemann Nielsen and Jensen, 1957; Sorokin and Kliashtorin, 1961). The rate of gross production was generally below 0.1 g C $\times$ m^{-2} $\times$ day^{-1}. Menzel and Ryther (1960) showed that the northern boundary of the Sargasso Sea proper is situated just south of the Bermuda Islands. The thermocline is permanent already a little south of the islands. This has the effect that the rate of production is low during the whole year. Off the Bermuda Islands, however, the thermocline breaks down during winter and early spring. During this time of the year nutrient salts are therefore introduced into the photic zone, giving rise to a rate of primary production of up to 0.9 g C $\times$ m^{-2} $\times$ day^{-1}.

During the international "Equalant" programme (cf., e.g., Zeitschel, 1969; Corcoran and Mahnken, 1969), intensive investigations were carried out in the tropical part of the Atlantic both during summer and winter. As was to be expected, near to Africa the variation in production rate was rather

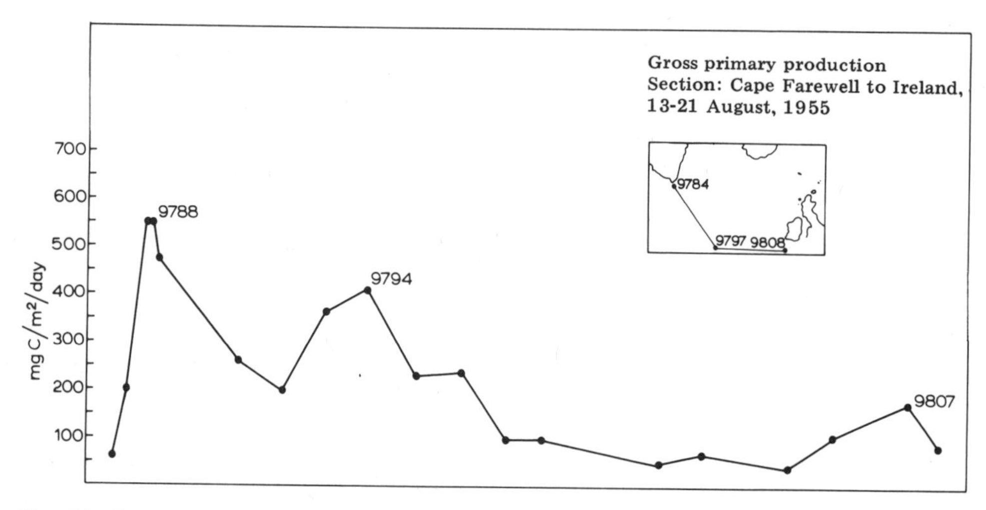

Fig. 59. Rate of gross primary production on a section from Cape Farewell (Greenland) to Ireland in August (after Hansen, 1959).

considerable. This is in agreement with the hydrographical conditions, which vary from one season to another. The rate of production decreased going from east to west.

Many observations are made in the open waters of the Atlantic at higher latitudes. An annual series was made on the continental shelf off New York, 41° N (Ryther and Yentsch, 1958).

In the northeastern part of the Atlantic, Hansen (1959) has made measurements of primary production during several years, unfortunately, however, only during the summer season. Regions with relatively low and regions with relatively high production values were found, the latter at places where currents with different water masses met producing a divergence, or where upwelling of nutrient-rich deep water occurs due to the presence of a submarine ridge. In Fig. 59 the curve shows the rate of primary production on a section from Cape Farewell (Greenland) to Ireland, in August. The maximum at Station 9788 is found in the border area between the East Greenland Polar Current and the Irminger Current. The maximum at Station 9794 is found over the Reykjanes Ridge. Finally, the maximum at Station 9807 is situated over the slope off Ireland.

In the western part of the South Atlantic only few measurements of the rate of primary production have been made at medium and higher latitudes. In the chart originally presented by Koblentz-Mishke et al. (1970) covering the annual rate of primary production in all world oceans (see Fig. 58), the Atlantic is characterized by a wealth of details. The only exception is the major part of the central eddy in the South Atlantic. The information for this area, although given with reservations, is most likely correct, however.

The adjacent waters of the Atlantic

For most of these basins quite detailed information is available. Starting with the Caribbean Sea and the Gulf of Mexico, we can state that the conditions for phytoplankton production vary greatly from one place to another. This is due to the very complicated hydrography of these waters, as upwelling is found at many places; thus, off the coast of Venezuela in the Caribbean Sea and at the shelf, both in the northern and the southwestern part of the Mexican Gulf, high rates of production are found there, such as are shown by, e.g., Curl (1960), Hammer (1967) and several Russian workers. In the central part of the Gulf of Mexico, Steele (1964), on the other hand, found rather low production rates. The same was shown for the central part of the Caribbean by Steemann Nielsen and Jensen (1957), Curl (1960) and Kabanova (1972). According to the latter, the annual rate of primary production there is only 40 g C/m^2.

The Mediterranean Sea is hydrographically characterized by the special interchange of water with the Atlantic through the Strait of Gibraltar. The influx of water takes place as a surface current, whereas the effluence takes place as a subsurface current. This has the effect that the Mediterranean looses nutrient salts due to the water exchange. Moreover, the conditions for primary production decrease from west to east. This is clearly shown by means of Fig. 60, where the two curves show the monthly rates of primary

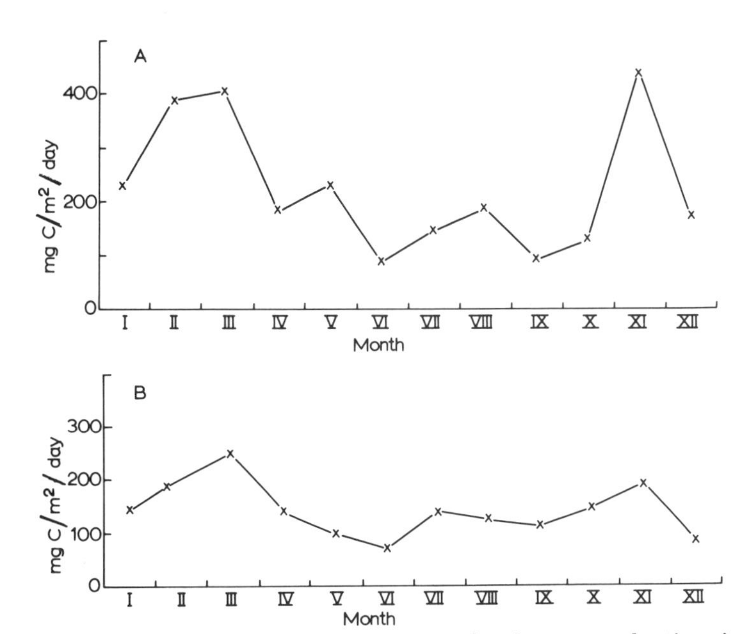

Fig. 60. The average monthly rates of primary production in the Mediterranean. A. 25—40 km off Barcelona, 1965—1967. B. Off Tira (Israel). (After Steemann Nielsen et al., 1969.)

production partly off the coast of Catalonia (Spain) in the west, partly from the coast of Israel in the east. In the open sea south of France, but also at the coast near Algiers, upwelling areas of limited extent are found during spring, giving rise to relatively high rates of primary production (e.g., Minas, 1969; Tellai, 1969).

In the North Sea and the Baltic, as well as in the waters in between, numerous series of primary productivity measurements have been made. Steele has published several articles (e.g., Steele, 1958) showing annual rates of primary production around 70 g C m^{-2} in the northern North Sea. The major part of production takes place during April—May.

Fig. 61 shows a curve presenting the monthly rates of gross production in the eastern part of the Kattegat obtained during seven years, when measurements during the whole time were made twice each month. The annual rate was about 100 g C m^{-2} (Steemann Nielsen, 1964a). Due to the special hydrographic conditions — a pronounced halocline — in years without ice-cover, the production takes place in all months of the year. From March to November the rate is relatively constant, only showing a slight maximum in July—September. This somewhat special production scheme for an area at 57° N is due to the rather constant intermixing from below, into the photic layer, of relatively nutrient-rich North Sea water. In the Great Belt, similar measurements have been made (Fig. 62). Although the annual rate of production is similar to that found in the eastern part of the Kattegat, the production scheme is somewhat different. The water masses of the photic layer in this strait have possibilities of contact with the bottom over many shallows.

The open parts of the Baltic must be characterized as relatively oligotrophic (cf., e.g., Sen Gupta, 1972). On the other hand, many of the coastal areas are very productive, especially those where large rivers debouch into more or less sheltered "Haffs" and bights.

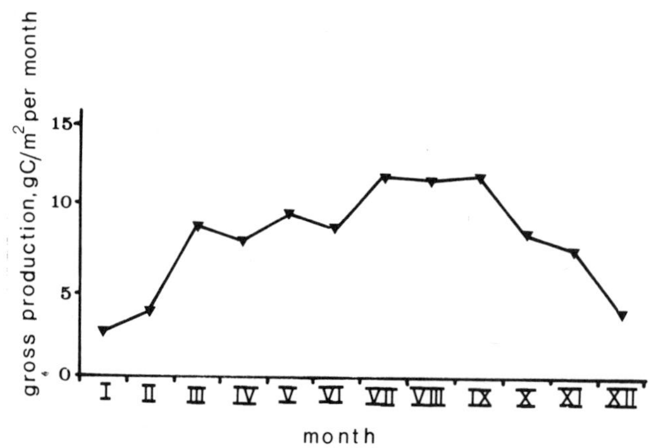

Fig. 61. The average monthly rates of gross production per m^2 in the Kattegat, Anholt Nord, 1954—1960 (after Steemann Nielsen, 1964a, revised).

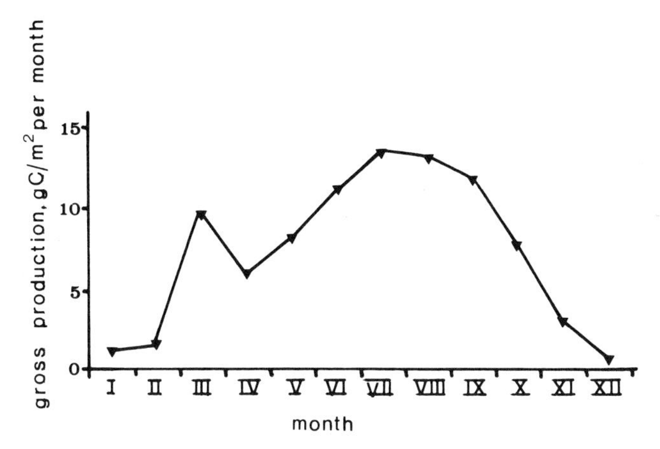

Fig. 62. The average monthly rates of gross production in the Great Belt, Halsskov Rev, 1953—1957 (after Steemann Nielsen, 1964a, revised).

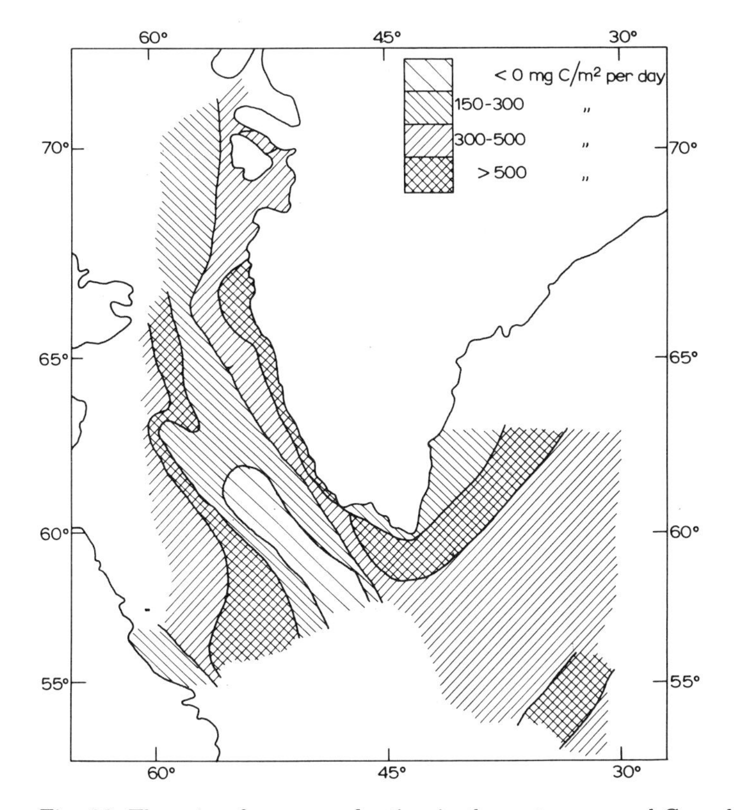

Fig. 63. The rate of gross production in the waters around Greenland, July—August, 1954 (after Steemann Nielsen, 1958, revised).

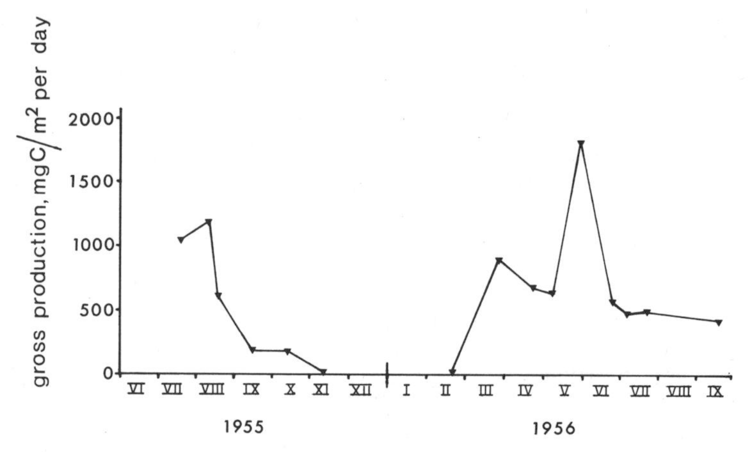

Fig. 64. The rate of gross production at the mouth of Godthaab Fjord, Greenland (after Steemann Nielsen, 1958, revised).

The adjacent seas to the north of the Atlantic have mostly been investigated concerning the rate of primary production during summer. Berge (1958) presented a close network of stations taken during the end of May in the Norwegian Sea. It was possible to show a very characteristic pattern for the spring maximum. Steemann Nielsen (1958) presented a survey of the gross production in the waters round Greenland during July—August (Fig. 63). High rates of production were found at all the localities where fronts between two current systems are found. This is due to the formation of eddies bringing up nutrient-rich water to the surface. It is found, e.g., at the front between the Labrador Current and the oceanic water in the centre of the Davis Strait. Very productive water is also found outside the coast of western Greenland between latitudes of about 62°N and 67°N. Over the banks there deeper water ascends toward the surface, thus enriching the photic layer with nutrients. Fig. 64 presents a curve showing the rate of primary production outside the Godthaab Fjord throughout nearly two years.

The Indian Ocean

Before the days of the International Indian Ocean Expeditions, which took place during the sixties, this ocean could be characterized as being much less investigated with respect to all sectors of oceanography than the other oceans. This was also the case concerning primary production. It was only during the Danish Galathea expedition (1950—52) that measurements first were made — especially in the tropics.

The chart in Fig. 58 (according to Koblentz-Mishke et al., 1970), which also covers the Indian Ocean, shows that the southern part of the ocean in principle, although by no means completely, corresponds rather well with the South Atlantic concerning rates of primary production. North to the Equator, on the other hand, conspicuous differences are found between the two oceans. This is due to the presence of the land masses in the Indian Ocean already at latitudes 10°—20° N. These have pronounced effects on the hydrographic conditions and therefore also on primary production. During the International Indian Ocean Expeditions, intensive work on the rate of primary production was carried out by many nations, e.g. in the western part by the U.S.A. (Ryther et al., 1966), Union of South Africa (Mitchel-Innes, 1967) and France (Angot, 1964). In the central, eastern and northern parts of the ocean, the measurements of primary production were made by the U.S.S.R. (Kabanova, 1968), India (Nair et al., 1968) and the U.S.A. (Ryther et al., 1966). The measurements from the southeastern Indian Ocean are published by El-Sayed and Jitts (1973).

North of the Equator, very high production rates were found off the coast of Somalia, in the Arabian Sea, and off the south coast of India. Due to the monsoon shifts, however, the rates of production vary greatly from one season to another. A central area with low rates of production is also found north of the Equator. It even penetrates as far as to the west coast of India, north of the area with high production. In the southern part of the Indian Ocean, areas with high production rates are found on the shelf off South Africa and in upwelling areas, e.g. southwest of Australia and south of Java. The central part of the ocean south of the Equator, forming an anticyclonic eddy like that found in the South Atlantic, is characterized by low production rates.

According to Prasad et al. (1970), the average rate of gross production is 0.24 g C $\times$ m^{-2} $\times$ day^{-1} in the western part of the Indian Ocean, whereas it is 0.19 in the eastern part. These rates seem to be somewhat higher than those found in the other oceans.

The Pacific Ocean

The Pacific Ocean has an area 3.5 times larger than that of the Indian Ocean and 1.7 times larger than that of the Atlantic. Very likely due to the enormous area, the hydrographical conditions in some areas are a little more complicated than those in the Atlantic. The main hydrographic trends, however, are the same as those in the Atlantic. Both in the Northern and the Southern Hemisphere, very large central parts constitute anticyclonic eddies where the surface water is sinking and where the rate of photosynthesis is therefore low.

Ryther et al. (1971) measured the rate of primary production in the upwelling area of the Peru coastal current. As to be expected, the rate in this region was extremely high, up to more than 10 g C $\times$ m^{-2} $\times$ day^{-1}, the highest rates ever measured in the sea. Rather high production rates were

TABLE IX

Area and annual rate of primary production in the different water masses of the Pacific (according to Koblentz-Mishke, 1967).

	Area (10^3 km^2)	Annual rate of production (g C × m^{-2})
Oligotrophic waters from the anticyclonic eddies	90,106	28
Transitional waters	33,358	49
Waters from the divergences near the Equator and from subpolar, oceanic regions	31,319	91
Waters off the coasts	20,423	105
Neritic waters	244	237

found off the Washington and Oregon coasts, where Anderson (1964) made investigations throughout the year.

The northwestern part of the Pacific Ocean is the best-known area of the Pacific concerning primary production. Important contributions are due both to Japanese and Russian workers (e.g., Saijo and Ichimura, 1960; Aruga and Monsi, 1962; and Taniguchi, 1972). Especially high production rates were found near the subarctic front, whereas the Kuroshio Current was characterized by medium rates, and the subtropical waters were poor concerning production. Jitts (e.g. 1965) has published measurements from the southwestern part of the Pacific near Australia. Both relatively productive and unproductive areas are found there.

Finally, Koblentz-Mishke (1967) has averaged the rates of production in the various parts of the Pacific. Table IX presents an extract.

As seen from Table IX, most of the water masses of the Pacific are oligotrophic.

Antarctic waters have only been relatively little investigated. Besides some few Russian articles, we may mention, amongst others, Bunt and Lee (1970), El-Sayed (1970) and El-Sayed and Jitts (1973). Although relatively high rates may be found in some places during some few summer months, the annual rates generally must be relatively low.

Ice is found in the North Polar Sea during the whole year, although with small channels of water between the ice floes during summer. Rates of primary production are therefore always extremely small (English, 1959).

MACROBENTHIC PLANTS

Compared with the wealth of investigations carried out concerning planktonic primary production, only little has been reported on benthic primary production, which includes that due to both microbenthos and macrobenthos.

Whereas Ryther (1963) estimated the global benthic production to be at least one tenth of the phytoplankton production, Bogorov (1969) arrived at a benthic production of only 0.5% of the phytoplankton production. This large discrepancy seems primarily to be due to an error by a factor of 10 made by Ryther in his calculation.

It is very difficult to estimate the production of benthic algae and sea grasses more than very approximately. Blinks (1955) presented rates of production of the algae off the Californian coast (about 37° N). They were measured both by means of weighing the standing stock and by experimentally determining the rate of photosynthesis. He gave values varying between 3 and 66 g dry weight $\times$ m^{-2} $\times$ day^{-1} corresponding to 0.1—22 g C. He referred to the few estimates made earlier in other parts of the world where maximum values had been found similar to those from Californian waters. Aleem (1973) measured the standing crop of *Macrocystis* in a kelp bed at La Jolla, California. He found that for *Macrocystis*, the standing crop varied between 6 and 10 kg fresh weight per m^2 during spring. To this should be added the weight of all the algae forming the understory of the kelp canopy. It was, on average, 4.8 kg per m^2. If these biomasses are computed as dry weight, the values are more or less the same as the maxima found more to the north off the Californian coast.

Odum and Odum (1955) estimated the daily rate of gross production to be about 12 g C $\times$ m^{-2} $\times$ day^{-1} on a coral reef in the tropical Pacific. About the same was found by Odum (1957) in vegetations of the marine turtle grass (*Thalassia*) in Florida. Recent measurements are presented by Ziemann (1973).

Grøntved (1958) estimated the total annual yield of dry matter by the macrobenthos in two shallow Danish fjords to be 201 and 184 g $\times$ m^{-2} (corresponding to about 70 and 60 g C), which figures are very likely underestimations. Mann (1972) has presented a table which shows that the production of the macrophytes ranges from 50 to 2,000 g C $\times$ $year^{-1}$ $\times$ m^{-2}.

It seems to be possible for the benthic macro-vegetation to attain higher production rates per surface area than the planktonic algae. This is very likely due to the scattering and "dilution" of the light in a vegetation of macrophytes, making it possible for the plants to utilize the light better than the planktonic algae.

MICROBENTHIC PLANTS

The first estimate of production of microbenthic algae was published by Pomeroy (1959). He showed that annual gross productivity in the salt marshes of Georgia was about 200 g C $\times$ m^{-2}. Using another technique (cf.

p. 96) Grøntved (1960) estimated the production in some sheltered, shallow fjords in Denmark to be 116 g C $\times$ m^{-2} $\times$ $year^{-1}$. Grøntved (1962) presented rather similar results from the Wadden Sea. Pamatmat (1968) and Gargas (1970) have published results from other areas which were of the same order of magnitude. Finally, Steele and Baird (1968) measured the production of microbenthic plants in a sandy bay on the Atlantic coast of Scotland. The effects of wave action at the habitat keep the population at a relatively low level, so that the yearly primary production of the benthic microflora is in the range 4—9 g C $\times$ m^{-2}.

Unpublished measurements in the Kattegat due to V. Hansen have shown that the rate of the microbenthic production can be estimated out to depths where, during summer, 5% of the green surface light is found. The production rate of the microbenthos in the Kattegat during the middle of summer, at depths between 12 and 20 m, was about 45% of that of the phytoplankton.

CHAPTER 15

MARINE PHOTOSYNTHESIS AND MAN

THE FOOD WEB IN THE SEA

Steemann Nielsen and Jensen (1957) estimated the annual net carbon productivity of all seas to be $12–15 \cdot 10^9$ tonnes. This is rather close to the newly published value by Koblentz-Mishke et al. (1970), which is $15–18 \cdot 10^9$ tonnes C. Although we cannot consider such estimates too rigorously — they are most likely somewhat too low — we must mention that they are very near to those estimated for all land areas on our globe.

Homo sapiens is, like all other animals, ultimately dependent on the organic matter produced by the photoautotrophic plants. The organic production in the world oceans is, as mentioned above, of the same order of magnitude as that on land. However, only about 1% of the human food originates at present from the sea — 5%, if we only consider proteins. There is no reason to believe that a high percentage ever will be possible. Several causes add to this apparently pessimistic view.

If man wants to get the maximum amount of food out of a certain area, he has to eat the primary producers directly. Products such as cabbage are, on land, the most effective. Already grain and potatoes are somewhat less effective products. Here we do not eat the organic matter produced by photosynthesis directly, but we utilize the organic matter which has been stored by the plants from one year to another.

However, the effectiveness of the products realized due to photosynthesis is reduced by a factor of about 10 if instead we eat the next trophic level, e.g. cattle. Where the food produced on land is concerned, in addition to the primary producers thus, we also eat the second trophic link, which we may call the herbivores. On the other hand, we do not eat the third link, the terrestrial carnivores such as foxes and lions. If man were forced to eat such animals exclusively, the human population would be extremely small. If man were to be able to utilize the food from the sea as effectively as that from the land, he would therefore first of all have to eat the primary producers directly. However, this is not possible. The main primary producers in the sea — the planktonic algae — cannot be harvested. In the oceans it would be necessary to filter about 50 m^3 water below each m^2 of the surface at least once a week. This is an impossible task. At present, we cannot even filter the next normal trophic link in the ocean — the herbivorous zooplankton — effectively, although this perhaps may be possible in the future in some areas of the sea, especially in the Antarctic, where a major part of the herbivores — the euphausids — are large. The giant whales are able to filter these euphausids effectively, man should be able to do the same.

Some fish — particularly sardines — are able to graze the phytoplankton directly. Such fish make up an important part of the world fish catch. Unfortunately, however, most of the exploited marine fish belong to the higher trophic links, that is to say to the "lions" of the sea. In shallow areas though, we do eat herbivores such as mussels.

If the present level of exploitation of the oceans in terms of carbon is taken into consideration, the total yield works out to be only 0.03% of the net primary production (Prasad et al., 1970). According to Steemann Nielsen and Jensen (1957), in eutrophic areas at higher latitudes, e.g. in the North Sea, about 0.2—0.3% of the carbon annually fixed by the phytoplankton is taken every year by the fishermen. With improved methods of fishing, the optimum seems to be at 0.4%.

Schaefer (1965) has attempted to estimate the potential yield of the sea. He considers — perhaps too optimistically — a production of $200 \cdot 10^6$ tonnes of fish for the world oceans as reasonable. That would mean a fourfold increase from the present level of exploitation, but would still only mean a yield of about 0.1% of the carbon assimilated by the phytoplankton.

In shallow water, a yield much higher than that for fish may be found, for example in beds of oysters and other mussels. These animals utilize the planktonic algae directly. In terms of world production, this kind of food production is only of small significance. Locally, however, it can be really important.

PRODUCTS OF MACROALGAE

Levring et al. (1969) presented an exhaustive review of all the products of macroalgae utilized in some way or other.

Products of marine macroalgae for human consumption have been known for many centuries in Japan, China, Korea and on many of the islands in the Pacific Ocean. In Japan it is still a substantial economic factor. Although the amount of consumable organic matter is of little importance compared with the number of the inhabitants of Japan, the so-called "Nori", which is made from species of *Porphyra* (red algae), is much esteemed. It is used as a vegetable or salad and is now quite expensive.

Many other algae, red, brown and green algae are eaten throughout the world, however, without adding much to the nutrition of the population of our globe.

Many macroalgae are used for making products of considerable technical importance; amongst them we mention agar, which is made from many red algae found throughout the world. It was originally used in the Far East as a gel-forming extract for cooking. It has now many other applications, one of which is in the manufacture of media for culturing bacteria.

Carrageenan is another important product made from red algae, in particular from *Chondrus crispus* and *Gigartina stellata*. There are numerous possibilities of applications, e.g. in the food industry (bakery, ice cream, etc.), for the clarification of beer and in the pharmaceutical industry and textile industries.

Alginic acid is a typical constituent of brown algae. The most important alga used for alginate production is *Macrocystis pyrifera.* This is due to its enormous size and the presence of large populations. Other species of brown algae are also utilized throughout the world; in Great Britain, for example, *Laminaria hyperborea* is used.

Alginic acid has a multitude of applications in many industries. It must now be considered to be by far the most important of all algal products used in industry.

REFERENCES

Aleem, A.A., 1973. Ecology of a kelp-bed in Southern California. *Bot. Mar.*, 16: 38—95.

Anderson, G.C., 1964. The seasonal and geographic distribution of primary productivity off the Washington and Oregon coasts. *Limnol. Oceanogr.*, 9: 284—302.

Angot, M., 1964. Production primaire de la region de Nosy-Bé, Août à Novembre 1963. *Cahiers O.R.S.T.O.M., Oceanogr.*, 2: 27—54.

Arens, K., 1936. Physiologisch polarisierter Massenaustausch und Photosynthese bei submersen Wasserpflanzen. II. Die $Ca(HCO_3)_2$-Assimilation. *Jahrb. Wiss. Bot.*, 83: 513—660.

Aruga, Y., 1965. Ecological studies of photosynthesis and matter production of phytoplankton. II. Photosynthesis of algae in relation to light intensity and temperature. *Bot. Mag. Tokyo*, 78: 360—365.

Aruga, Y. and Monsi, M., 1962. Primary production in the north-western part of the Pacific off Honshu, Japan. *J. Oceanogr. Soc. Japan*, 18: 85—94.

Atkins, W.R.G., Jenkins, P.G. and Warren, F.J., 1954. The suspended matter in sea water and its seasonal changes as affecting the visual range of the Secchi disc. *J. Mar. Biol. Assoc. U.K.*, 33: 497—509.

Becacos-Kontos, T. and Svansson, A., 1969. Relation between primary production and irradiance. *Mar. Biol.*, 2: 140—144.

Berge, G., 1958. The primary production in the Norwegian Sea in June 1954, measured by an adapted ^{14}C technique. *Rapp. P.-V. Réun. Cons. Perm. Int. Explor. Mer*, 144: 85—91.

Biebl, R., 1952. Resistenz der Meeresalgen gegen sichtbares Licht und gegen kurzwellige UV-Strahlen. *Protoplasma*, 41: 353—377.

Blinks, L.R., 1955. Photosynthesis and productivity of littoral marine algae. *J. Mar. Res.*, 14: 363—373.

Bogorov, V.G., 1969. Productivity of the ocean; primary production and its utilization in food chains. In: *Int. Oceanogr. Congr., Moscow, 1966, Unesco, 2nd*, pp. 117—124.

Boysen Jensen, P., 1918. Studies on the production of matter in light- and shadow-plants. *Dan. Bot. Tidskr.*, 36: 219—262.

Braarud, H. and Klem, A., 1931. Hydrographical and chemical investigations in the coastal waters off Möre and in the Romsdalsfjord. Hvalråd. *Skr. Nor. Vidensk. Akad. Oslo*, 1: 1—88.

Brown, A.H., 1953. The effects of light on respiration using isotopically enriched oxygen. *Am. J. Bot.*, 40: 719—729.

Buch, K., 1960. Kohlendioxyd im Meerwasser. *Encycl. Plant Physiol.*, 5 (part 1): 46—61.

Bunt, I.S. and Lee, E.C., 1970. Seasonal primary production in antarctic sea ice at McMurdo Sound in 1967. *J. Mar. Res.*, 28: 304—320.

Burnett, J.H., 1965. Functions of carotenoids other than in photosynthesis. In: T.W. Goodwin (Editor), *Chemistry and Biochemistry of Plant Pigments.* Academic Press, London, pp. 381—403.

Calvin, M., 1965. Chemical evolution. *Proc. R. Soc. Lond., Ser. A*, 288: 441—466.

Castenholz, R.W., 1964. The effect of daylight and light intensity on the growth of littoral marine diatoms in culture. *Physiol. Plant.*, 17: 951—963.
Corcoran, E.F. and Mahnken, C.V.N., 1969. Productivity of the tropical Atlantic Ocean. In: *Proc. Symp. Oceanic and Fish Research in the Tropical Atlantic, Abidjan, October, 1966*, pp. 57—68.
Curl, H., 1960. Primary production measurements in the northern coastal waters of South America. *Deep-Sea Res.*, 7: 183—189.

Dodge, J.D., 1973. *The Fine Structure of Algal Cells.* Academic Press, London, 261 pp.
Doty, M.S. and Oguri, M., 1957. Evidence for a photosynthetic daily periodicity. *Limnol. Oceanogr.*, 2: 37—40.
Duysens, L.N.M., 1951. Transfer of light energy within the pigment systems present in photosynthesizing cells. *Nature*, 168: 548—550.

Echlin, P., 1966. Origins of photosynthesis. *Sci. J.*, April 1966, 7 pp.
Egle, K., 1960. Menge und Verhältnis der Pigmente. *Encycl. Plant Physiol.*, 5 (1): 444—496.
El-Sayed, S.Z., 1970. On the productivity of the Southern Ocean. In: *Antarctic Ecology.* Academic Press, London, 1: 119—135.
El-Sayed, S.Z. and Jitts, H.R., 1973. Phytoplankton production in the southeastern Indian Ocean. In: B. Zeitschel (Editor), *The Biology of the Indian Ocean. Ecol. Stud., Anal. Synth.*, 3: 132—142.
Emerson, R. and Lewis, C.M., 1942. The photosynthetic efficiency of phycocyanin in *Chroococcus* and the problem of carotenoid participation in photosynthesis. *J. Gen. Physiol.*, 25: 579—595.
Emerson, R. and Lewis, C.M., 1943. The dependence of the quantum yield of *Chlorella* photosynthesis on wave length of light. *Am. J. Bot.*, 30: 165—178.
Encyclopedia of Plant Physiology, V, Parts 1 and 2, 1960. Springer Verlag, Berlin.
English, T.S., 1959. Primary production in the central North Polar Sea, drifting station Alpha, 1957—58. In: *Int. Oceanogr. Congr., New York, N.Y., 1959, 1st*, pp. 838—839.

Faurholt, C., 1924. *Studier over Kuldioxyd og Kulsyre og over Karbaminater og Karbonater.* Thesis, University of Copenhagen, 124 pp.
Fogg, G.E., 1968. *Photosynthesis.* English Universities Press, London, 116 pp.
Fogg, G.E., Nalewajko, C. and Watt, W.D., 1965. Extracellular products of phytoplankton photosynthesis. *Proc. R. Soc. Lond., Ser. B.*, 162: 517—534.
French, C.S., 1938. The rate of CO_2 assimilation by purple bacteria at various wave lengths of light. *J. Gen. Physiol.*, 21: 21—87.

Gaarder, T. and Gran, H.H., 1927. Investigations of the production of plankton in the Oslo Fjord. *Rapp. P.-V. Réun. Cons. Perm. Int. Explor. Mer.*, 42: 3.
Gabrielsen, E.K., 1948. Effects of different chlorophyll concentrations on photosynthesis in foliage leaves. *Physiol. Plant.*, 1: 5—37.
Gabrielsen, E.K., 1960a. Beleuchtungsstärke und Photosynthese. *Encycl. Plant Physiol.*, 5(part 2): 27—48.
Gabrielsen, E.K., 1960b. Lichtwellenlänge und Photosynthese. *Encycl. Plant Physiol.*, 5(part 2): 49—78.
Gaidukov, N., 1902. Über den Einfluss farbigen Lichts auf die Färbung lebender Oscillarien. *Abh. Preuss. Akad. Wiss., Berlin*, 5: 1—36.
Gargas, E., 1970. Measurements of primary production, dark fixation and vertical distribution of the microbenthic algae in the Øresund. *Ophelia*, 8: 231—253.

Gargas, E., 1971. "Sun-shade" adaptation in microbenthic algae from the Øresund. *Ophelia*, 9: 107—112.
Gessner, F., 1938. Die Beziehung zwischen Lichtintensität und Assimilation bei submersen Wasserpflanzen. *Jahrb. Wiss. Bot.*, 86: 491—526.
Gessner, F., 1944. Der Chlorophyllgehalt der Seen als Ausdruck ihrer Produktivität. *Arch. Hydrobiol.*, 40: 687—732.
Gessner, F., 1955. *Hydrobotanik. Die physiologischen Grundlagen der Pflanzenverbreitung im Wasser. I. Energiehaushalt.* VEB Deutscher Verlag der Wissenschaften, Berlin, 517 pp.
Grøntved, J., 1958. Underwater macrovegetation in shallow coastal waters. *J. Cons. Int. Explor. Mer.*, 24: 32—42.
Grøntved, J., 1960. On the productivity of microbenthos and phytoplankton in some Danish fjords. *Medd. Dan. Fisk. Havunders., N.S.*, 3: 55—92.
Grøntved, J., 1962. Preliminary report on the productivity of microbenthos and phytoplankton in the Wadden Sea. *Medd. Dan. Fisk. Havunders. N.S.*, 3: 347—378.

Halldal, P., 1968. Photosynthetic capacities and photosynthetic action spectra of endozoic algae of the massive coral *Favia*. *Biol. Bull.*, 134: 111—124.
Halldal, P., 1974. Light and photosynthesis of different marine algal groups. In: N.G. Jerlov and E. Steemann Nielsen (Editors), *Optical Aspects of Oceanography*. Academic Press, London, pp. 345—360.
Hammer, L., 1967. Die Primärproduktion im Golf von Cariace (Ost Venezuela). *Int. Rev. Ges. Hydrobiol.*, 52: 757—768.
Hammer, L., 1968. Salzgehalt und Photosynthese bei marinen Pflanzen. *Mar. Biol.*, 1: 185—190.
Hansen, V.K., 1959. Danish investigations on the primary production and the distribution of chlorophyll *a* at the surface of the North Atlantic during summer. *Int. Counc. Explor. Sea, Spec. IGY Meeting*, 5: 1—4.
Harder, R., 1923. Über die Bedeutung von Lichtintensität und Wellenlänge für die Assimilation farbiger Algen. *Z. Bot.*, 15: 305—355.
Harder, R., 1933. Über die Assimilation der Kohlensäure bei konstanten Aussenbedingungen. II. Das Verhalten von Sonnen- und Schattenpflanzen. *Planta*, 20: 699—733.
Hatch, M.D. and Slack, C.R., 1970. Photosynthetic CO_2-fixation pathways. *Ann. Rev. Plant Physiol.*, 21: 141—161.
Haxo, F.T. and Blinks, L.R., 1950. Photosynthetic action spectra of marine algae. *J. Gen. Physiol.*, 33: 389—422.
Hentschel, E., 1933—1936. Allgemeine Biologie des Südatlantischen Ozeans. *Wiss. Ergeb. Dtsch. Atl. Exped. Forsch. Vermessungssch. "Meteor" 1925—1927*, 9: 1—343.
Hood, D.W. and Park, K., 1963. Bicarbonate utilization by marine phyto-plankton in photosynthesis. *Physiol. Plant*, 15: 273—282.
Hunding, C. and Hargrave, B.T., 1973. A comparison of benthic microalgal production measured by C^{14} and oxygen methods. *J. Fish. Res. Board Can.*, 30: 309—312.
Hutchinson, G.E., 1957. *A Treatise on Limnology. I. Geography, Physics and Chemistry.* John Wiley, New York, N.Y., 1015 pp.

Ichimura, S., Saijo, Y. and Aruga, Y., 1962. Photosynthetic characteristics of marine phytoplankton and their ecological meaning in the chlorophyll method. *Bot. Mag. Tokyo*, 75: 212—220.

Jackson, W.A. and Volk, R.J., 1970. Photorespiration. *Ann. Rev. Plant Physiol.*, 21: 385—432.
Jerlov, N.G., 1951. Optical studies of ocean water. *Rep. Swed. Deep-Sea Exped.*, 3: 1—59.

Jerlov, N.G., 1968. *Optical Oceanography. Elsevier Oceanography Series 5.* Elsevier, Amsterdam, 194 pp.

Jerlov, N.G., 1970. General aspects of underwater daylight and definitions of fundamental concepts. In: E. Kinne (Editor), *Marine Ecology. Comprehensive, Integrated Treatise on Life in Oceans and Coastal Waters, 1. Part 1.* John Wiley, New York, N.Y., pp. 95—102.

Jerlov, N.G. and Nygaard, K., 1969. A quanta and energy meter for photosynthetic studies. *Inst. Fys. Oceanogr.*, 10: 1—19.

Jitts, H.R., 1963. The simulated in situ measurement of oceanic primary production. *Aust. J. Mar. Freshwater Res.*, 14: 139—147.

Jitts, H.R., 1965. The summer characteristics of primary productivity the Tasman and Coral Seas. *Aust. J. Mar. Freshwater Res.*, 16: 151—162.

Johnson, N.G. and Levring, T., 1946. The photosynthetical effect of ultra-violet radiation on some marine algae. *Medd. Oceanogr. Inst. Göteborg,* 11: 1—5.

Jørgensen, E.G., 1964. Adaptation to different light intensities in the diatom *Cyclotella meneghiniana* Kütz. *Physiol. Plant,* 17: 136—145.

Jørgensen, E.G., 1966. Photosynthetic activity during the life cycle of synchronous *Skeletonema* cells. *Physiol. Plant.*, 19: 789—799.

Jørgensen, E.G., 1968. The adaptation of plankton algae. II. Aspects of the temperature adaptation of *Skeletonema costatum. Physiol. Plant.*, 21: 423—427.

Jørgensen, E.G., 1969. The adaptation of plankton algae. IV. Light adaptation in different algal species. *Physiol. Plant.*, 22: 1307—1315.

Jørgensen, E.G. and Steemann Nielsen, E., 1965. Adaptation in plankton algae. *Mem. Ist. Ital. Idrobiol.*, 18(suppl.): 37—46.

Kabanova, Y.G., 1968. Primary production in the northern Indian Ocean. *Okeanologiya,* 8: 270—278 (in Russian).

Kabanova, Y.G., 1972. Dependence of primary production values upon different factors in the northeastern part of the Caribbean Sea. *Okeanologiya,* 12: 299—314 (in Russian).

Kalle, K., 1966. The problem of the Gelbstoff in the sea. *Oceanogr. Mar. Biol. Ann. Rev.*, 4: 91—104.

Karekar, M.D. and Joshi, G.V., 1973. Photosynthetic carbon metabolism in marine algae. *Bot. Mar.*, 16: 216—220.

Kimbal, H.H., 1935. Intensity of solar radiation at the surface of the earth and its variation with latitude, altitude, season and time of the day. *Mon. Weather Rev.*, 63: 1.

Kniep, H., 1914. Über die Assimilation und Atmung der Meeresalgen. *Int. Rev. Ges. Hydrobiol. Hydrograph.*, 7: 1—38.

Koblentz-Mishke, O.J., 1967. Primary production. In: V.G. Kort (Editor), *Tychyi okean. Biolog. Tychogo okeana. 1. Plankton.* Nauka, Moscow, 267 pp.

Koblentz-Mishke, O.J., Volkovinsky, V.V. and Kabanova, J.G., 1970. Plankton primary production of the world ocean. In: *Scientific Exploration of the South Pacific.* Natl. Acad. Sci., Washington, D.C., Standard Book No. 309-01755-6., pp. 84—94.

Krinsky, N.J., 1964. Carotenoid de-epoxidations in algae. I. Photochemical transformation of autheraxanthin to zeaxanthin. *Biochim. Biophys. Acta,* 88: 487—491.

Krinsky, N.J., 1966. The role of carotenoid pigment as protective agents against photosensitized oxidation in chloroplasts. In: T.W. Goodwin (Editor), *Biochemistry of Chloroplasts,* 1. Academic Press, London, pp. 423—430.

Lampe, H., 1935. Die Temperatureinstellung der Stoffgewinnung bei Meeresalgen als plasmatische Anpassung. *Protoplasma*, 23: 534—578.

Levring, T., 1947. Submarine daylight and the photosynthesis of marine algae. *Göteborg K. Vetensk. Vitterh.- Samh. Handl.*, 5(6): 1—90.

Levring, T., Hoppe, H.A. and Schmid, O.J., 1969. *Marine Algae. A Survey of Research and Utilization*, 1. Botanica Marina Handbooks, De Gruyter, Hamburg, 421 pp.

Lewis, J.R., 1964. *The Ecology of Rocky Shores*. The English Universities Press, London, 323 pp.

Lohmann, H., 1908. Untersuchungen zur Feststellung des vollständigen Gehaltes des Meeres an Plankton. *Wiss. Meeresunters. N.F., Abt. Kiel*, 10: 129—370.

Lohmann, H., 1920. Die Bevölkerung des Ozeans mit Plankton nach den Zentrifugenfänge während der Ausreise der "Deutschland" 1911. Zugleich ein Beitrag zur Biologie des Atlantischen Ozeans. *Arch. Biontolog.*, 4: 1—617.

Loomis, W.E., 1960. Historical introduction (photosynthesis). *Encycl. Plant Physiol.*, 5(part 1): 85—114.

Lorenzen, C.J., 1972. Extinction of light in the ocean by phytoplankton. *J. Cons. Int. Explor. Mer.*, 34: 262—267.

Lowenhaupt, B., 1956. The transport of calcium and other cations in submerged aquatic plants. *Biol. Rev.*, 31: 371—395.

Lund, S., 1959. The marine algae of East Greenland. II. Geographic distribution. *Medd. Grønl.*, 156: 1—72.

Mann, K.H., 1972. Macrophyte production and detritus food chains in coastal waters. *Mem. Ist. Ital. Idrobiol.*, 29(Suppl.): 353—383.

McFarland, W.N. and Prescott, J., 1959. Standing crop, chlorophyll content and in situ metabolism of a giant kelp community in southern California. *Publ. Inst. Mar. Sci. Univ. Texas*, 6: 109—132.

McLeod, G.C., 1957. The effect of circularly polarized light on the photosynthesis and chlorophyll *a* synthesis of certain marine algae. *Limnol. Oceanogr.*, 2: 360—362.

Menzel, D.W. and Ryther, J.H., 1960. The annual cycle of primary production in the Sargasso Sea off Bermuda. *Deep-Sea Res.*, 6: 351—367.

Minas, H.J., 1969. Rapport sur les résultats de recherches portant sur les mésures de la production organique primaire en certains secteurs du Bassin Méditerranéen Nord-Occidental (Golfe de Marseille et Mer de Ligurie en particulier). *NATO Subcomm. Oceanogr. Tech. Rep.*, No. 47. Mediterr. Productivity Project, pp. 24—34.

Mitchell-Innes, B.A., 1967. Primary production studies in the south-west Indian Ocean 1961—1963. *S. Afr. Assoc. Mar. Biol. Res., Oceanogr. Res. Inst. Invest. Rep.*, 14: 1—20.

Montfort, C., 1930. Die photosynthetischen Leistungen roter Tiefseealgaen und Grottenflorideen an freier Meeresoberfläche. *Protoplasma*, 19: 385—413.

Montfort, C., 1931. Assimilation und Stoffgewinn der Meeresalgen bei Aussüssung und Rückversalzung. I. Phasen der Giftwirkung und die Frage der Reversibilität. II. Typen der funktionellen Salzeinstellung. *Ber. Dtsch. Bot. Ges.*, 49: 49—66.

Montfort, C., 1936. Umwelt, Erbgut und physiologische Gestalt. I. Lichttod und Starklichtresistung bei Assimilationsgeweben. *Jahrb. Wiss. Bot.*, 84: 1—57.

Morris, J. and Glover, H.E., 1974. Questions on the mechanism of temperature adaptation in marine phytoplankton. *Mar. Biol.*, 24: 147—154.

Myers, J., 1946. Culture conditions and the development of the photosynthetic mechanism. IV. Influence of light intensity on photosynthetic characteristics of *Chlorella*. *J. Gen. Physiol.*, 29: 429—440.

Nair, P.V.R., Samuel, S., Joseph, K.J. and Balachandran, V.K., 1968. Primary production and potential fishery resources in the sea around India. In: *Symp. "Living Resources in the Sea around India", Cochin, 1968, ICAR* (unpublished but cited according to Prasad et al., 1970).

Odum, H.T., 1957. Primary production measurements in eleven Florida springs and a marine turtle grass community. *Limnol. Oceanogr.*, 2: 85—97.
Odum, H.T. and Odum, E.P., 1955. Trophic structure and productivity at a windward coral reef community on Eniwetok Atoll. *Ecol. Monogr.*, 25: 291—320.
Österlind, S., 1949. Growth conditions of the alga *Scenedesmus quadricauda*, with special reference to the inorganic carbon sources. *Symb. Bot. Upps.*, 10: 1—141.

Paasche, E., 1964. A tracer study of the inorganic carbon uptake during coccolith formation and photosynthesis in the coccolithophorid *Coccolithus huxleyi*. *Physiol. Plant.*, 3 (Suppl.): 1—82.
Pamatmat, M.M., 1968. Ecology and metabolism of a benthic community on an intertidal sandflat. *Int. Rev. Ges. Hydrobiol.*, 53: 211—298.
Patten, B.C., 1968. Mathematical models of plankton production. *Int. Rev. Ges. Hydrobiol.*, 53: 357—408.
Pechlaner, R., 1971. Factors that control the production rate and biomass of phytoplankton in high-mountain lakes. *Mitt. Int. Ver. Limnol.*, 19: 125—145.
Petersen, C.G.J., 1913. *Om Baendeltangens (Zostera marina) Aarsproduktion i de Danske Farvande.* (Mindeskrift for Japetus Steenstrup.) Bianco Luno, Copenhagen, 20 pp.
Pirson, A., 1957. Induced periodicity of photosynthesis and respiration in *Hydrodictyon*. In: H. Gaffron et al. (Editors), *Research in Photosynthesis.* Interscience Press, New York, N.Y., pp. 490—499.
Pirson, A. and Lorenzen, H., 1966. Synchronized dividing algae. *Ann. Rev. Plant Physiol.*, 17: 439—458.
Platt, T., 1969. The concept of energy efficiency in primary production. *Limnol. Oceanogr.*, 14: 653—659.
Pomeroy, L.R., 1959. Algal productivity in salt marshes of Georgia. *Limnol. Oceanogr.*, 4: 386—397.
Prasad, R.R., Banerji, S.K. and Nair, P.V.R., 1970. A quantitative assessment of the potential fishery resources of the Indian Ocean and adjoining seas. *Indian J. Anim. Sci.*, 40: 73—98.

Rabinowitch, E., 1951. *Photosynthesis and Related Processes,* 2. Part 1. Interscience, New York, N.Y., 599 pp.
Raven, J.A., 1970. Exogenous inorganic carbon sources in plant photosynthesis. *Biol. Rev.*, 45: 167—221.
Romose, V., 1940. Ökologische Untersuchungen über *Homalothecium sericeum*, Seine Wachstumsperioden und seine Stoffproduktion. *Dan. Bot. Ark.*, 10: 1—134.
Ruttner, F., 1926. Über den Gaswechsel von *Elodea*-Sprossen verschiedener Tiefenstandorte under den Lichtbedingungen grösserer Seetiefen. *Planta,* 2: 588—599.
Ryther, J.H., 1963. Geographic variations in productivity. In: M.N. Hill (Editor), *The Sea, Ideas and Observations on Progress in the Study of the Seas, 2.* Interscience, New York, N.Y., pp. 347—380.
Ryther, J.H. and Menzel, D.W., 1959. Light adaptation by marine phytoplankton. *Limnol. Oceanogr.*, 4: 492—497.
Ryther, J.H. and Yentsch, C.S., 1958. Primary production of continental shelf waters off New York. *Limnol. Oceanogr.*, 3: 327—335.
Ryther, J.H., Hall, J.R., Pease, A.K., Bakun, A. and Jones, M.M., 1966. Primary organic production in relation to the chemistry and hydrography of the western Indian Ocean. *Limnol. Oceanogr.*, 11: 107—113.
Ryther, J.H., Menzel, D.W., Hilburt, E.M., Lorenzen, C.J. and Corvin, N., 1971. The production and utilization of organic matter in the Peru coastal current. *Inv. Pesq.*, 35: 43—59.

Saijo, Y. and Ichimura, S., 1960. Primary production in the northwestern Pacific Ocean. *J. Oceanogr. Soc. Japan*, 16: 29—145.
Saijo, Y. and Ichimura, S., 1962. Some considerations on photosynthesis of phytoplankton from the point of view of productivity measurement. *J. Oceanogr. Soc. Japan*, 20: 687—692.
Schaefer, M.B., 1965. The potential harvest from the sea. *Trans. Am. Fish. Soc.*, 94: 123—128.
Schindler, D.W., Schmidt, R.V. and Reid, R.A., 1972. Acidification and bubbling as an alternative to filtration in determining phytoplankton production by the ^{14}C method. *J. Fish. Res. Board Can.*, 29: 1627—1631.
Schott, G., 1935. *Geographie des Indischen und Stillen Ozeans.* C. Boysen, Hamburg, 423 pp.
Schott, G., 1942. *Geographie des Atlantischen Ozeans.* 3. Aufl. Hamburg, 368 pp.
Sen Gupta, R., 1972. Photosynthetic production and its regulating factors in the Baltic Sea. *Mar. Biol.*, 17: 82—92.
Seybold, A. and Weissweiler, A., 1942. Spektrophotometrische Messungen an grünen Pflanzen und an Chlorophyll-Lösungen. *Bot. Arch.*, 43: 252—290.
Sharp, J.H. and Renger, E.H., 1973. Extracellular production of organic matter by marine phytoplankton. *Inst. Mar. Resour. Progr. Rep. July 72—June 73*, pp. 69—72.
Smayda, T.J. and Mitchell-Innes, B., 1974. Dark survival of autotrophic, planktonic marine diatoms. *Mar. Biol.*, 25: 195—202.
Sorokin, C., 1957. Changes in photosynthetic activity in the course of cell development in *Chlorella. Physiol. Plant.*, 10: 659—666.
Sorokin, C. and Kraus, R.W., 1959. Maximum growth rates of *Chlorella* in steady-state and in synchronized cultures. *Proc. Natl. Acad. Sci. U.S.*, 45: 1740—1744.
Sorokin, Y.J. and Kliashtorin, 1961. Primary production in the Atlantic Ocean. *Tr. Vses. Gidrobiol. O.-Va. Akad. Nauk S.S.S.R.*, 11: 265—284 (in Russian).
Sournia, A., 1969. Cycle annuel du phytoplankton et de la production primaire dans les mers tropicales. *Mar. Biol.*, 3: 287—303.
Stålfelt, M.G., 1960. Vorleben, Aktivierung, Inaktivierung, "Ermüdung", Wundreiz. *Encycl. Plant. Physiol.*, 5(part 2): 186—212.
Steele, J.H., 1958. Production studies in the northern North Sea. *Rapp. P.-V. Réun. Cons. Perm. Int. Explor. Mer*, 144: 79—84.
Steele, J.H., 1962. Environmental control of photosynthesis in the sea. *Limnol. Oceanogr.*, 7: 137—150.
Steele, J.H., 1964. A study of the production in the Gulf of Mexico. *J. Mar. Res.*, 22: 211—220.
Steele, J.H. and Baird, I.E., 1968. Production biology of a sandy beach. *Limnol. Oceanogr.*, 13: 14—25.
Steemann Nielsen, E., 1935. The production of phytoplankton at the Faroe Isles, Iceland, East Greenland and in the waters around. *Medd. Komm. Dan. Fisk. Havunders. Plankton*, 3: 1—93.
Steemann Nielsen, E., 1947. Photosynthesis of aquatic plants with special reference to the carbon-sources. *Dan. Bot. Ark.*, 12(8): 1—71.
Steemann Nielsen, E., 1952. The use of radio-active carbon (C^{14}) for measuring organic production in the sea. *J. Cons. Int. Explor. Mer.*, 18: 117—140.
Steemann Nielsen, E., 1954. On the preference of some freshwater plants in Finland for brackish water. *Bot. Tidskr.*, 51: 242—247.
Steemann Nielsen, E., 1955. The interaction of photosynthesis and respiration and its importance for the determination of ^{14}C-discrimination in photosynthesis. *Physiol. Plant*, 8: 945—953.
Steemann Nielsen, E., 1958. A survey of recent Danish measurements of the organic productivity in the sea. *Rapp. P.-V. Réun. Cons. Perm. Int. Explor. Mer.*, 144: 92—95.

Steemann Nielsen, E., 1960. Uptake of CO_2 by the plant. *Encyclop. Plant Physiol.*, 5(part 1): 70—84.

Steemann Nielsen, E., 1961. Chlorophyll concentration and rate of photosynthesis in *Chlorella vulgaris. Physiol. Plant.*, 14: 868—876.

Steemann Nielsen, E., 1962. Inactivation of the photochemical mechanism in photosynthesis as a means to protect cells against too high light intensities. *Physiol. Plant.*, 15: 161—171.

Steemann Nielsen, E., 1963. On bicarbonate utilization by marine phytoplankton in photosynthesis, with a note on carbamino carboxylic acids as a carbon source. *Physiol. Plant.*, 16: 466—469.

Steemann Nielsen, E., 1964a. Investigations of the rate of primary production at two Danish light ships in the transition area between the North Sea and the Baltic. *Medd. Dan. Fisker. Havunders. N.S.*, 4: 31—77.

Steemann Nielsen, E., 1964b. Recent advances in measuring and understanding marine primary production. *J. Ecol.*, 52(Suppl.): 119—130.

Steemann Nielsen, E., 1964c. On a complication in marine productivity work due to the influence of ultraviolet light. *J. Cons. Int. Explor. Mer.*, 39: 130—135.

Steemann Nielsen, E., 1965. On the determination of the activity in ^{14}C-ampoules for measuring primary production. *Limnol. Oceanogr.*, 10(Suppl.): R247—R252.

Steemann Nielsen, E. and Hansen, V.K., 1959. Light adaptation in marine phytoplankton populations and its interrelation with temperature. *Physiol. Plant.*, 12: 353—370.

Steemann Nielsen, E. and Hansen, V.K., 1961. Influence of surface illumination on plankton photosynthesis in Danish waters (56°N) throughout the year. *Physiol. Plant.*, 14: 595—613.

Steemann Nielsen, E. and Jensen, E.A., 1957. Primary oceanic production. The autotrophic production of organic matter in the oceans. *Galathea Rep.*, 1: 49—135.

Steemann Nielsen, E. and Jørgensen, E.G., 1968a. The adaptation of plankton algae. I. General Part. *Physiol. Plant.*, 21: 401—413.

Steemann Nielsen, E. and Jørgensen, E.G., 1968b. The adaptation of plankton algae. III. With special consideration of the importance in nature. *Physiol. Plant.*, 21: 647—654.

Steemann Nielsen, E. and Park, S.T., 1964. On the time course in adapting to low light intensities in marine phytoplankton. *J. Cons. Int. Explor. Mer*, 29: 19—24.

Steemann Nielsen, E. and Willemoës, M., 1971. How to measure the illumination rate when investigating the rate of photosynthesis of unicellular algae under various light conditions. *Int. Rev. Ges. Hydrobiol.*, 56: 541—556.

Steemann Nielsen, E. and Wium-Andersen, S., 1971. The influence of Cu on photosynthesis and growth in diatoms. *Physiol. Plant.*, 24: 480—484.

Steemann Nielsen, E. and Wium-Andersen, S., 1972. Influence of copper on photosynthesis of diatoms, with special reference to an afternoon depression. *Verh. Int. Verein. Limnol.*, 18: 78—83.

Steemann Nielsen, E., Hansen, V.K. and Jørgensen, E.G., 1962. The adaptation to different light intensities in *Chlorella vulgaris* and the time dependence on transfer to a new light intensity. *Physiol. Plant.*, 15: 505—517.

Steemann Nielsen, E., Battaglia, B. and Minas, H.J. (Editors), 1969. Mediterranean productivity project. *NATO Subcommittee on Oceanographic Research, Tech. Rep.*, No. 47, 102 pp.

Steemann Nielsen, E., Wium-Andersen, S. and Rochon, T., 1974. On problems in G.M. countings in the C^{14}-technique. In: *Congr. Int. Assoc. Limnol., Winnipeg, 1974*, p. 198.

Strickland, J.D.H., 1958. Solar radiation penetrating the ocean. A review of requirements, data and methods of measurement, with particular reference to photosynthetic productivity. *J. Fish. Res. Board Can.*, 15: 453—493.

Strickland, J.D.H., 1960. Measuring the production of marine phytoplankton. *Fish. Res. Board Can. Bull.*, 122: 1—172.

Sverdrup, H.N., 1953. On conditions for the vernal blooming of phytoplankton. *J. Cons. Int. Explor. Mer.*, 18: 287—295.

Sverdrup, H.N., 1955. The place of physical oceanography in oceanographic research. *J. Mar. Res.*, 14: 287—294.

Talling, J.F., 1957. The phytoplankton population as a compound photosynthetic system. *New. Phytol.*, 56: 133—149.

Talling, J.F., 1966. Photosynthetic behaviour in stratified and unstratified lake populations of a planktonic diatom. *J. Ecol.*, 54: 99—127.

Talling, J.F., Wood, R.B., Prosser, M.V. and Baxter, R.M., 1973. The upper limit of photosynthetic productivity by phytoplankton: evidence from Ethiopian soda lakes. *Freshwater Biol.*, 3: 53—76.

Tamiya, H., 1957. Mass cultures of algae. *Ann. Rev. Plant Physiol.*, 8: 309—334.

Tanada, T., 1951. The photosynthetic efficiency of carotenoid pigments in *Navicule minima*. *Am. J. Bot.*, 38: 276—283.

Taniguchi, A., 1972. Geographical variation of primary production in the western Pacific Ocean and adjacent seas with reference to the inter-relations between various parameters of primary production. *Mem. Fac. Fish. Hokkaido Univ.*, 19: 1—33.

Tellai, S., 1969. Détermination de la production organique à l'aide du carbon 14, dans les parages d'Alger. *NATO Subcomm. Oceanogr. Res. Tech. Rep.*, No. 47. Mediterr. Productivity Project, pp. 56—63.

Tseng, C.K. and Sweeney, B.M., 1946. Physiological studies of *Gelidium cartilagineum*. I. Photosynthesis, with special reference to the carbon dioxide factor. *Am. J. Bot.*, 33: 706—715.

Vollenweider, R.A. (Editor), 1974. *A Manual on Methods for Measuring Primary Production in Aquatic Environments.* IBP Handbook, No. 12. Blackwell, Oxford, 2nd ed., 213 pp.

Watt, W.D., 1965. Release of dissolved organic material from the cells of phytoplankton populations. *Proc. R. Soc. Lond., Ser. B.*, 164: 521—551.

Watt, W.D. and Paasche, E., 1963. An investigation of the conditions for distinguishing between CO_2 and bicarbonate utilization in algae according to the methods of Hood and Park. *Physiol. Plant.*, 16: 674—681.

Watt, W.D. and Fogg, G.E., 1966. The kinetics of extracellular glycollate production by *Chlorella pyrenoidosa.* J. Exp. Bot., 17: 117—134.

Weller, S. and Franck, J., 1941. Photosynthesis in flashing light. *J. Phys. Chem.*, 45: 1359—1373.

Westlake, D.F., 1965. Some basic data for investigations of the productivity. *Mem. Ist. Ital. Idrobiol.*, 18(Suppl.): 229—248.

Wetzel, R.G., 1964. A comparative study of primary productivity of higher aquatic plants, periphyton and phytoplankton in a large, shallow lake. *Int. Rev. Ges. Hydrobiol.*, 49: 1—61.

Wetzel, R.G., 1965. Techniques and problems of primary productivity measurements in higher aquatic plants and periphyton. *Mem. Ist. Ital. Idrobiol.*, 18(Suppl.): 249—267.

Wilce, R.T., 1967. Heterotrophy in arctic sublittoral seaweeds: an hypothesis. *Bot. Mar.*, 10: 185—197.

Winokur, M., 1948. Photosynthesis relationships of *Chlorella* species. *Am. J. Bot.*, 35: 207—217.

Yentsch, C.S. and Ryther, J.H., 1957. Short-term variations in phytoplankton chlorophyll and their significance. *Limnol. Oceanogr.*, 2: 140—142.
Yocum, C.S. and Blinks, L.R., 1954. Photosynthetic efficiency of marine plants. *J. Gen. Physiol.*, 38: 1—16.
Yocum, C.S. and Blinks, L.R., 1958. Light-induced efficiency and pigment alterations in red algae. *J. Gen. Physiol.*, 41: 1113—1117.

Zeitschel, B., 1969. Productivity and microbiomass in the tropical Atlantic in relation to the hydrographical conditions (with emphasis on the eastern area). *Proc. Symp. Ocean Fish. Res. Trop. Atlantic, Abidjan, October 1966*, pp. 69—84.
Zieman, J.C., 1973. Quantitative and dynamic aspects of the ecology of turtle grass, *Thalassia testudinum*. Preprint from: *Recent Adv. Estuarine Res., Proc. Int. Estuarine Res. Conf., 2nd, Myrtle Beach, S.C., 1973.*

INDEX